CHEMICAL GRAPH THEORY

Mathematical Chemistry

A series of books edited by:
Danail Bonchev, Department of Physical Chemistry, Higher Institute of Chemical Technology, Burgas, Bulgaria
Dennis H. Rouvray, Department of Chemistry, University of Georgia, Athens, USA

Volume 1
CHEMICAL GRAPH THEORY: Introduction and Fundamentals

Volume 2
CHEMICAL GRAPH THEORY: Reactivity and Kinetics

Additional volume in preparation

Volume 3
CHEMICAL GROUP THEORY: Introduction and Fundamentals

This book is part of a series. The publisher will accept continuation orders which may be cancelled at any time and which provide for automatic billing and shipping of each title in the series upon publication. Please write for details.

CHEMICAL GRAPH THEORY
Reactivity and Kinetics

Edited by

Danail Bonchev

*Department of Physical Chemistry, Higher Institute
of Chemical Technology, Burgas, Bulgaria*

and

Dennis H. Rouvray

*Department of Chemistry, University of Georgia
Athens, USA*

ABACUS PRESS

an imprint of

GORDON AND BREACH SCIENCE PUBLISHERS
Philadelphia Reading Paris Montreux Tokyo Melbourne

Gordon and Breach Science Publishers

5301 Tacony Street, Drawer 330
Philadelphia, Pennsylvania 19137
United States of America

Post Office Box 90
Reading, Berkshire RG1 8JL
United Kingdom

58, rue Lhomond
75005 Paris
France

Post Office Box 161
1820 Montreux 2
Switzerland

3-14-9, Okubo
Shinjuku-ku, Tokyo 169
Japan

Private Bag 8
Camberwell, Victoria 3124
Australia

Library of Congress Cataloging-in-Publication Data

Chemical graph theory: reactivity and kinetics / edited by Danail
 Bonchev and Dennis H. Rouvray.
      p.   cm. -- (Mathematical chemistry; v. 2)
   Includes bibliographical references and index.
   ISBN 0-85626-515-2
   1. Chemistry--Mathematics.   2. Graph theory.   I. Bonchev, Danail.
II. Rouvray, D. H.   III. Series.
QD39.3.G73C49   1992
540'. 1'5115--dc20   92-6825
                                                                       CIP

CONTENTS

INTRODUCTION TO THE SERIES

The mathematization of chemistry has a long and colorful history extending back well over two centuries. At any period in the development of chemistry the extent of the mathematization process roughly parallels the progress of chemistry as a whole. Thus, in 1786 the German philosopher Immanuel Kant observed[1] that the chemistry of his day could not qualify as one of the natural sciences because of its insufficient degree of mathematization. It was not until almost a century later that the process really began to take hold. In 1874 one of the great pioneers of chemical structure theory, Alexander Crum Brown (1838–1922), prophesied[2] that "...chemistry will become a branch of applied mathematics; but it will not cease to be an experimental science. Mathematics may enable us retrospectively to justify results obtained by experiment, may point out useful lines of research and even sometimes predict entirely novel discoveries. We do not know when the change will take place, or whether it will be gradual or sudden...." This prophecy was soon to be fulfilled. Indeed, even before these words were uttered, combinatorial methods were being employed for the enumeration of isomeric species.[3] During Crum Brown's lifetime algebraic equations were used to predict the properties of substances, calculus was employed in the description of thermodynamic and kinetic behavior of chemical systems, and graph theory was adapted for the structural characterization of molecular species.

In the present century the applications of mathematics have come thick and fast. The advent of quantum chemistry in the 1920s brought in its wake a host of mathematical disciplines that chemists felt obliged to master. These included several areas of linear algebra, such as matrix theory and group theory, as well as calculus. Group theory has become so widely accepted by chemists that it is now used routinely in areas such as crystallography and molecular structure analysis. Graph theory seems to be following in the footsteps of group theory and is currently being exploited in a wide range of applications involving the classification, systemization, enumeration and design of systems of chemical interest. Topology has found important applications in areas as diverse as the characterization of potential energy surfaces, the discussion of chirality, and the description of catenated and knotted molecular species. Information theory has yielded valuable insights into the nature of thermodynamic processes and the origin of life. The contemporary fascination with dissipative systems, fractal phenomena and chaotic behavior has again introduced new mathematics, such as catastrophe theory and fractal geometry, to the chemist.

All of these and numerous other applications of mathematics that have been made in the chemical domain have brought us to a point where we consider it may now be fairly said that mathematics plays an indispensable role in modern chemistry. Because of the burgeoning use of mathematics by chemists and the current feeling that mathematics is opening up some very exciting new directions to explore, we believe that the 1990s represent a particularly auspicious time to present a comprehensive treatment of the manifold applications of mathematics to chemistry. We were persuaded to undertake this somewhat awesome task after much reflection and eventually decided to publish our material in a series of volumes, each of which is to be devoted to a discussion of the applications of a specific branch of mathematics. The title of our series, *Mathematical Chemistry*, was chosen to reflect as accurately as possible the proposed contents. The term "mathematical chemistry" was coined in the early 1980s to designate the field that concerns itself with the novel and nontrivial application of mathematics to chemistry. Following the usual practice in this area, we shall interpret chemistry very broadly to include not only the traditional disciplines of inorganic, organic and physical chemistry but also its hybrid offspring such as chemical physics and biochemistry.

It is anticipated that each of the volumes in our series will contain five to six separate chapters, each of which will be authored by a leading expert in the respective field. Whenever it is evident that one such volume is insufficient to do justice to a wealth of subject matter, additional volumes devoted to applications of the same branch of mathematics will be published. In this way it is hoped that our coverage will indeed be comprehensive and reflect significant developments made up to the end of the twentieth century. Our aim will be not only to provide a background survey of the various areas we cover but also to discuss important current issues and problems, and perhaps point to some of the major trends that might be reasonably expected in mathematical chemistry in the early part of the new millennium. In the first few volumes of our series we propose to examine the applications to chemistry of graph theory, group theory, topology, combinatorics, information theory and artificial intelligence.

It may be of interest to observe here that mathematical chemists have often applied and even sought after branches of mathematics that have tended to be overlooked by the chemical community at large. This is not to imply that the mathematics itself is necessarily new — in fact, it may be quite old. What is new is the application to chemistry; this is why the word novel was employed in our earlier definition of mathematical chemistry. The thrill of discovering and developing some novel application in this sense has been an important source of motivation for many mathematical chemists. The other adjective used in our definition of mathematical chemistry, i.e., nontrivial, is also worthy of brief comment. To yield profitable new insights, the mathematics exploited in a chemical context usually needs to be of at least a reasonably high level. In an endeavor to maintain a uniformly high level, we shall seek to ensure that all of the contributions to our volumes are written by researchers at the forefront of their respective disciplines. As a consequence, the contents of our various volumes are likely to appeal to a fairly sophisticated audience: bright undergraduate and postgraduate students, researchers operating at the tertiary level in academia, industry or government service, and perhaps even to newcomers to the area desirous of experiencing an invigorating excursion through the realms of mathematical chemistry. Overall, we hope that our series will provide a valuable resource for scientists and mathematicians seeking an authoritative and detailed account of mathematical techniques to chemistry.

In conclusion, we would like to take this opportunity of thanking all our authors, both those who have contributed chapters so far and those who have agreed to submit contributions for forthcoming volumes. It is our sincere hope that the material to be presented in our series will find resonance with our readership and afford many hours of enjoyable and stimulating reading.

Danail Bonchev
Dennis H. Rouvray

1. I. Kant, *Metaphysiche Anfangsgründe der Naturwissenschaft*, Hartknoch Verlag, Riga, 1786.
2. A. Crum Brown, *Rept. Brit. Assoc. Adv. Sci.*, 45–50, 1874.
3. F.M. Flavitsky, *J. Russ. Chem. Soc.*, **3**, 160, 1871.

PREFACE

This is the second volume to appear in our ongoing series of monographs entitled *Mathematical Chemistry*. As indicated in volume 1 of our series, each monograph will be devoted to discussion of the major applications of mathematics in various areas of chemistry. Volume 2, like its predecessor, focuses on the chemical applications of graph theory. In fact, the present volume complements the first by elaborating on a number of applications that could not be addressed in volume 1 for reasons of limited space. Having provided a suitable graph-theoretical background in volume 1, we move on here to a detailed coverage of the role of graph theory in the study of chemical kinetics, reaction mechanisms and quantitative structure-activity relationships (QSAR). An important message that is conveyed by the authors of this volume is that graph theory is a very valuable and powerful tool in the hands of the theoretical chemist.

In chapter 1 Petrov discusses how graph-theoretical principles can be applied to study the kinetics of heterogeneous catalytic reactions. In his discussion special attention is paid to gas-phase reactions that take place over a solid catalyst; a number of results are included that have not hitherto been published. Chapter 2 by Temkin and Bonchev presents a very comprehensive treatment of the classification and coding of reaction mechanisms based on graph theory methods. Having dealt with the classification of mechanisms, Plath, Hass and Kramer in chapter 3 proceed to discuss the mechanistic description of chemical processes with special reference to pericyclic reactions and some remarks on aromaticity. In chapter 4 Nikanorov and Sokolov explore operator networks that can be employed to interpret revolutionary inter-relationships between the various chemical entities involved in chemical reactions. Among other things, it is demonstrated by these authors that many well-known chemical regularities have their origins in various forms of similarity and may be incorporated into the framework of the operator-network approach to chemical reactions. The concluding chapter by Mercier, Sobel and Dubois offers a review of the DARC/PELCO method that has been used successfully for many years as a tool in QSAR searches and describes a number of recent innovations in the use of the method for the design of pharmaceutical drugs.

Once again, we wish to express our gratitude to so many of our colleagues — too numerous to mention individually by name — who are currently active within the expanding field of chemical graph theory. Without their continual encouragement and support our own participation in this field over the past two decades would not have been nearly as rewarding and fulfilling. Finally, we hope that some of the excitement and stimulation that we have felt over the years may be experienced by our readers and perhaps lead them to become contributors to this burgeoning field.

Danail Bonchev
Dennis H. Rouvray

Chapter 1

APPLICATION OF GRAPH THEORY TO STUDY OF THE KINETICS OF HETEROGENEOUS CATALYTIC REACTIONS

L.A. Petrov

Institute of Kinetics and Catalysis, Bulgarian Academy of Sciences
ul. acad. G. Bonchev bl. 11, Sofia 1113, Bulgaria

1.1 Introduction

Graph theory has been widely applied in recent years in studies on chemical kinetics. It has proved to be a powerful tool in the theoretical investigation of reaction mechanisms, kinetic modeling and especially for the derivation of kinetic equations for chemical processes. King and Altman (1) adopted for discussion of the kinetics of enzymic reactions a method whose basis and formalisms are graph-theoretical in nature. Temkin (2) applied graph-theoretical ideas to kinetic studies of heterogeneous catalytic reactions with the aim of clarifying the structure and mechanism of the independent routes of complex catalytic reactions. However, the possibilities of graph theory methods have not been fully exploited in his approach. The methods of graph theory were used for the first time in kinetics studies in the works of Vol'kenstein and Gol'dstein (3, 4). Recently, two reviews on the applications of graph theory in chemical kinetics have also been published (5, 6).

In this chapter we summarize results on the application of graph theory to the kinetics of heterogeneous catalytic reactions published up to the present. Special attention is paid to gas-phase reactions over a solid catalyst. A number of more recent results, which have hitherto not been published, are also included.

1.2 Graph Theory Definitions

A number of monographs (7–9) in which the current state-of-the-art in graph theory is set forth have appeared. A short introduction to the basic ideas and definitions of graph theory has been presented in Volume 1 of this series of books. We give here only a few definitions to be used in our further exposition.

The graph $G(X, Y)$ is a manifold which consists of two subsets: a nonzero subset X, the elements of which are called vertices, and a subset Y, the elements of which are called edges. Each edge is defined by two vertices. If the edge has an assigned direction, it is referred to as arc. Each arc of a graph is characterized by its weight. Graphs which consist of arcs are said to be oriented. In this chapter we shall use only oriented graphs.

The cycle is a chain of the graph consisting of arcs and vertices which begins and finishes on the same vertex. One of most important characteristics of any graph is its cyclomatic number Ω which gives the number of independent cycles it contains. This number can be determined from the following formula:

$$\Omega = s - n + 1, \tag{1}$$

where s is the number of arcs in the graph, and n is the number of vertices in the graph.

A graph without cycles is described as a tree. Trees which consist of all the vertices of a graph connected with arcs are called spanning (maximal) trees. To construct a spanning tree one should delete from the graph Ω arcs. The weight of the spanning tree is equal to the product of the weights of its arcs. If the spanning tree has several branches, the weight of the spanning tree is equal to the sum of the weights of the branches. The weight of a branch equals the product of the weights of its arcs. Weights of the parallel arcs between two vertices are additive.

For convenience, the vertices of the graph are denoted by a number. Then any graph or its subgraph can be represented in two possible ways:

(a) By means of the vertices incorporated in them. For example, some of the cycles of the graph shown in Fig. 1.1 are represented as follows:

1. 2. 3

1. 2

2. 3

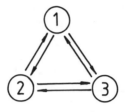

Figure 1.1

(b) By means of the arcs incorporated in them. The same cycles then are represented as follows:

12. 23. 31
12. 21
23. 32

Both methods are equivalent and use is made of the appropriate one in any given case.

1.3 Application of Graph Theory to Study of the Kinetics of Heterogeneous Catalytic Reactions

The application of graph theory to the study of the kinetics of heterogeneous catalytic reactions involves the following stages:
1. Construction of the supposed mechanism of the process.
2. Construction of the kinetic graph and evaluation of the weight of each of its arcs.
3. Determination of the basic steps for each route.
4. Determination of the base determinants for each vertex of the graph.
5. Determination of the cyclic characteristics of each route.
6. Determination of the conjugation parameter for each route.
7. Construction of a kinetic equation.
8. Analysis of the constructed model.
   The first three stages are not involved directly with the methods of the graph theory. At all these stages, the researcher makes use of existing information on the mechanism of the process and the nature of the catalyst obtained from kinetic, chemical, and

physical methods. It is still difficult to formalize this work, but a number of efforts in this direction have already been made. Such efforts entail determination of the set of reaction mechanisms and their classification by the methods of topological analysis (10, 11), group theory (12), Hougen–Watson (13), and Yang and Hougen (14); all are based on analysis of the nature of the absorbed substances and the possible limiting steps, and make use of the methods of graph theory (15, 16, 17). A separate chapter in this book discusses the ideas advanced in (16) and (17) concerning the classification of linear mechanisms.

1.4 Elements of the Kinetics of Complex Heterogeneous Catalytic Reactions

We shall characterize as a complex heterogeneous catalytic reaction any reaction system which consists of at least of two independent stoichiometric equations. The compounds participating in a heterogeneous catalytic reaction can be divided into two groups, viz. reagents and intermediate surface compounds (ISCs). The regents are the starting compounds and products of the reaction. Their concentrations or partial pressures can be determined at any moment of the reaction and at any point in the reaction space.

The stoichiometry and structure of the ISCs are not exactly known. In most cases, the concentrations of the ISCs cannot be determined directly and so they are expressed by concentrations or partial pressures of the reagents via adsorption isotherms.

Every heterogeneous catalytic reaction proceeds via a certain number of elementary reactions. Elementary reactions are reactions in which only one energetic barrier is overcome. The stiochiometric coefficients of elementary reactions are integers. The rate of an elementary reaction obeys the Law of Mass Action – or the Law of Surface Action if the reactions take place at the surface of a catalyst. The Arrhenius equation is strictly valid only for elementary reactions.

Two elementary reactions proceeding in forward and reverse directions form the elementary step. Three types of elementary steps are defined: mass transfer steps, adsorption–desorption steps, and surface reaction steps. In this chapter we shall deal only with latter two types of elementary steps. If the elementary reaction is irreversible, the elementary step consists of one elementary reaction. Participants in elementary reactions are the ISCs or the ISCs together with reagents. An elementary reaction whose rate depends linearly on the concentration of the ISCs will be called a linear elementary reaction.

The set of elementary reactions that allows a qualitative and quantitative description of major characteristics of the process studied to be made, will be termed the mechanism of the chemical reaction. A mechanism consisting only of linear steps will be described as a linear mechanism. In our further exposition we shall study only linear mechanisms although a large number of reactions proceed via a nonlinear mechanism, e.g. they include elementary reactions having rates that depend nonlinearly on the concentration of the ISCs. These classes of mechanisms can be formally described within the framework of a linear model if the assumption is made that the nonlinear steps are equilibria which proceed at a high rate so that it is possible to combine them with slow linear steps.

We discuss here the method described in (15) for the application of graph theory in construction of reaction mechanisms. As stated before, all substances participating in a heterogeneous catalytic reaction may be divided into two groups: substances which do not contain the catalyst (reagents) and substances containing the catalyst (ISCs). Let the reagents, whose composition is always known, consist of just two chemical elements, A and B. The chemical formula of each of them will be $A_q B_m$, where the set U of the pairs of integers (q, m) is assumed known. The chemical formula of the intermediates will be $A_i B_j K$, where K is the catalyst, and pairs of the integers (i, j) belong to the set V. It is normal to assume that set U is a subset of V. The authors of (15) assumed the opposite and proposed the following constraints on i and j:

$$q_{min} \leqslant i \leqslant q_{max}$$
$$m_{min} \leqslant j \leqslant m_{max}, \tag{2}$$

which are valid only for isomerization or decomposition reactions of the starting compound. This, of course, restricts the field of application and the number of mechanisms that can be generated. The set V should icnlude the pair $(0, 0)$ corresponding to the free surface of the catalyst.

The elementary reaction between two intermediates, $A_i B_j K$ and $A_j' B_j' K$, is possible only if such reagents exist in the set that guarantees the material balance of the reaction. In other words, two vertices of the kinetic graph, $A_i B_j K$ and $A_i' B_j' K$, will be connected by an arc if the set of pairs contains pairs such that:

$$i + q = i' + q'$$
$$j + m = j' + m' \tag{3}$$

Taking into account the stoichiometric restrictions imposed on set V, one can construct the necessary graph with the help of (3). The mechanism of the reaction corresponding to this graph will include all possible steps permitted by the material balance. The imposed weak restrictions allow the construction of a kinetic graph which probably contains as a subgraph the real mechanism of the process. Differentiation between the possible mechanisms is based on our knowledge of the chemistry of the process.

Complex reactions can be expressed as a manifold of elementary steps and ISCs which in certain conditions give different reaction pathways while producing the same or different products. Such pathways are called reaction routes. For one completion of a given reaction route every step must participate as many times as its stoichiometric number. Summing stoichiometric equations of elementary steps in a given route multiplied by their stoichiometric numbers should yield the stoichiometric equition of the route. The reaction rate on a reaction route is equal to the number of realizations of the reaction route in unit time.

To illustrate the above ideas we now give examples of linear and nonlinear mechanisms. The example of a one route linear mechanism is that of the reaction of the configurational isomerization of 1,2-dimethyl-cyclohexane on a nickel catalyst (18):

1) cis-1, 2-$C_8H_{16} + Z = $ cis-1, 2-$C_8H_{16-2n}Z + nH_2$

2) cis-1, 2-$C_8H_{16-2n}Z = $ trans-1, 2-$C_8H_{16-2n}Z$ $\tag{4}$

3) trans-1, 2-$C_8H_{16-2n}Z + nH_2 = $ trans-1, 2-$C_8H_{16} + Z,$

where Z is a free site on the surface of the catalyst, and cis-1, 2-$C_8H_{16-2n}Z$ and trans-1, 2-$C_8H_{16-2n}Z$ are ISCs.

The mechanism of a two-route process of selective (I) and total (II) oxidation of ethylene on a silver catalyst is an example of nonlinear mechanism (19)

$$
\begin{array}{lll}
 & \text{I} & \text{II} \\
1) \quad Z + O_2 = ZO_2 & 2 & 0 \\
2) \quad ZO_2 + C_2H_4 \rightarrow ZO + C_2H_4O & 2 & 0 \\
3) \quad 2ZO = 2Z + O_2 & 1 & -3 \quad (5) \\
4) \quad C_2H_4 + 6ZO \rightarrow 2CO_2 + H_2O + 6Z & 0 & 1 \\
\hline
I) \quad 2C_2H_4 + O_2 = 2C_2H_4O & & \\
II) \quad C_2H_4 + 3O_2 = 2CO_2 + 2H_2O & &
\end{array}
$$

In the above mechanism steps 1) and 2) are linear and steps 3) and 4) are nonlinear.

The first route consists of steps 1), 2), and 3) with stoichiometric numbers $\sigma_1^{(1)} = 2$, $\sigma_2^{(1)} = 2$, and $\sigma_3^{(1)} = 1$. The second route consists of steps 3) and 4) with stoichiometric numbers $\sigma_3^{(2)} = -3$ and $\sigma_4^{(2)} = 1$. The third elementary step participates in both routes with different stoichiometric numbers. The minus sign for $\sigma_3^{(2)}$ means that elementary step 3) in the second route is proceeding in the reverse direction. On the right side of the mechanism (5) are given the stoichiometric numbers for both routes. At the end of mechanism (5) are given stoichiometric equations for both routes.

The rate of the sth elementary reaction connecting the intermediates α and β

$$
\sum_{j=1}^{m} \mu_{js} C_j + X_\alpha = \sum_{i=1}^{m'} \mu_{is} C_i + X_\beta \tag{6}
$$

may therefore be expressed as:

$$
r_s = k_s X_\alpha \prod_{j=1}^{m} C_j^{\mu_{js}}, \tag{7}
$$

where r_s is the rate of the sth elementary reaction, and X_α and X_β are the concentrations of the ISCs α and β involved in the sth step,

C_j and C_i are the concentrations of the jth and ith reagents participating in the sth elementary step, μ_{js} and μ_{is} are the stoichiometric coefficients of the jth and ith reagents participating in the sth elementary step, m and m' are the number of reagents taking part in the forward and reverse elementary reactions of the sth step.

Existing experimental data show that in most cases in elementary reactions μ_{js} and $\mu_{is} = 0$ or 1, and m and $m' = 0$ or 1. When this holds, equation (6) can be expressed as:

$$C_j + X_\alpha = C_i + X_\beta \text{ or } X_\alpha \rightarrow X_\beta \tag{8}$$

and equation (7) will take the form:

$$r_s^L = k_s^L X_\alpha C_j \quad \text{or}$$
$$r_s^L = k_s^L X_\alpha, \tag{9}$$

depending on whether the reagents participate in the sth elementary reaction or not.

If a elementary step s is part of one route mechanism, its rate r_s will be given by the difference in the rates of elementary reactions in the forward r_s^+ and reverse r_s^- directions:

$$r_s = r_s^+ - r_s^- = \sigma_s R, \tag{10}$$

where σ_s is the stoichiometric number of the elementary step s in the reaction mechanism, and R is the reaction rate for the route.

If the elementary step forms part of a P-route mechanism, the rate of the elementary step r_s will equal the sum of the rates along the different routes in which the elementary step participate, multiplied by the stoichiometric number $\sigma_s^{(P)}$ of the elementary step for the given route (20, 21):

$$r_s = r_s^+ - r_s^- = \sum_{p=1}^{P} \sigma_s^{(P)} R^{(P)}, \tag{11}$$

where $\sigma_s^{(P)}$ is the stoichiometric number of the elementary step s for the pth route, $R^{(P)}$ is the reaction rate along the pth route, and P is the number of routes, in which participate the elementary step s.

The system (11), derived by Temkin, containing the equations for the reaction rate of all elementary steps of given mechanism is referred to as the "steady-state condition". It represents the conditions necessary for the reaction to proceed in the steady state regime. By solving this system one can obtain expressions for the concentrations of the ISCs, which contain only concentrations or partial pressures of the reagents taking part in this reaction system.

The "steady-state condition" may be represented in terms of the concentrations of the ISCs (Bodenstein Rule) as follows:

$$X_t \sum_{r=1}^{n-1} \sigma_{tr} = \sum_{s=1}^{n-1} \sigma_{ts} X_s$$

$$\sum_{r=1}^{n-1} X_r = C, \tag{12}$$

where X_t is the concentration of the intermediate t, $\Sigma \sigma_{tr}$ is the sum of stoichiometric numbers of the elementary steps in which ISC X_t is consumed, $\Sigma \sigma_{ts} X_s$ is the sum in which each term is formed by a concentration of an ISC (which is transformed into Xt) multiplied by the stoichiometric number of the corresponding elementary step, n is the total number of the elementary steps in the mechanism in study, and C is the total concentration of the ISCs which is constant.

The maximum number of linearly independent chemical reactions, M_T, in a complex reaction system is given by:

$$M_T = N - q, \tag{13}$$

where N is the total number of reagents in the system, q is the rank of the atomic matrix A of the reagents. Each element of atomic matrix $[a_{ij}]$ gives the number of atoms of a given type which contain a certain reagent, $i = 1, 2, \ldots, N$, $j = 1, 2, \ldots, N_1$ (number of types of atoms).

The linearly independent chemical reactions create the stoichiometric (thermodynamic) basis for the reaction system. All other chemical processes in the system can be expressed as a linear function of these basic reactions. The stoichiometric basis is determined by the stoichiometry of the process, the number and composition of the reagents, e.g. by quantities which can be directly determined experimentally.

According to Horiuti, the number of independent routes M_K in a complex reaction system is given by the following expression:

$$M_K = S - L, \tag{14}$$

where S is the number of elementary steps in the reaction mechanism, and L is the number of independent ISCs taking part in the reaction mechanism.

Horiuti's rule is valid both for linear and nonlinear mechanisms, though we shall restrict our study to linear mechanisms.

The surface concentrations of the ISCs are interconnected by the following balance equation:

$$\Theta_0 + \sum_{i=1} \Theta_i = 1, \tag{15}$$

where Θ_0 is the free part of the catalyst surface, and Θ_i is that part of the catalyst surface occupied by the ith ISC. L_{tot} is the total number of ISCs.

The number of independent ISCs L will be equal to:

$$L = L_{tot} - 1. \tag{16}$$

We now substitute for L, using the corresponding expression from equation (16) in equation (14), and obtain the result:

$$M_K = S - L_{tot} + 1. \tag{17}$$

The basis given by equation (17) is called a kinetic basis. It consists of the routes obtained from the proposed reaction mechanism. Their number is different for different mechanisms and depends on the details of the mechanism.

It is obvious that

$$M_T \leqslant M_K, \tag{18}$$

which means that some of the routes of the kinetic basis are dependent from a thermodynamic point of view. In this case the reaction mechanism should be changed so that the thermodynamic and kinetic bases become equal. If the reaction mechanism preserves the number of elementary steps and ISC, the difference $M_K - M_T$ gives the number of so-called null routes. Such a route has a stoichiometric equation $0 = 0$ but its rate is different from zero.

The rate of the reaction for stoichiometrically independent routes (which we cannot measure experimentally) is calculated from the reaction rates for the individual reagents in the system (which can be measured directly using the experiment) using the equation:

$$r_i = \sum_{j=1}^{M_T} [\mu_{ij}]^T R_j, \tag{19}$$

where r_i is the reaction rate for ith reagent, $1 = 1, 2, \ldots, n$ is the number of the reagents in the system, R_j is the reaction rate along the stoichiometrically linearly independent route j, $j = 1, 2, \ldots$, M_T, $[\mu_{ij}]^T$ is an element of transposed stoichiometric matrix.

Each element $[\mu_{ij}]$ of the stoichiometric matrix gives the stoichiometric coefficient of the ith reagent in the jth reaction independent route.

1.5 Relationship Between a Graph and a Mechanism for a Heterogeneous Catalytic Reaction

Let us form two sets for each linear reaction mechanism. The elements X_1 of the first set correspond to the ISCs participating in the reaction mechanism. The elements Y_1 of the second set correspond to the elementary reactions of the reaction mechanism. From our earlier definition of a graph, follow that each linear mechanism can be represented by a graph, if the set X_1 of the mechanism containing the ISCs is isomorphic to the subset X of the graph containing the vertices of the graph, and the set Y_1 of the mechanism containing the elementary reactions is isomorphic to the subset Y of the graph containing the arcs of the graph. A graph that is isomorphic to a reaction mechanism will be called a kinetic graph. Each arc of the kinetic graph will correspond to an elementary reaction of the mechanism and be defined by two vertices, e.g. by two ISCs participating in the elementary reaction.

In the following, a kinetic graph should always be understood to be associated with its corresponding linear reaction mechanism and vice versa; a linear reaction mechanism should always be associated with its corresponding reaction graph.

The kinetic graph shown in Fig. 1.1 corresponds to mechanism (4) based on the isomorphism between the reaction mechanism and the graph. Here vertex 1 corresponds to the free site Z on the catalyst surface, vertex 2 corresponds to the ISC cis-1, 2-$C_8H_{16-2n}Z$, and vertex 3 corresponds to the ISC trans-1, 2-$C_8H_{16-2n}Z$. All steps in the mechanism (4) are reversible. For every elementary reaction of the mechanism in the kinetic graph separate arcs appear and for every elementary step in the reaction graph two arcs with the opposite directions appear.

Each arc connecting two intermediates has a weight that is determined by the Law of Mass Action or Law of Surface Action. The weight of an arc of a kinetic graph will be defined as the ratio between the rate of the elementary reaction and the concentration of the ISCs participating in the elementary reaction. The weights of the ISCs can thereby be expressed in terms of the concentrations or partial pressures of the reagents. Accordingly, the kinetic equations contain only terms which can be estimated experimentally. Hence, the weights w_s of the elementary reaction s in general case (using equation (7)) will be given by:

$$w_s = \frac{r_s}{X_s} = k_s \prod_{j=1}^{m} c_j^{\mu_{js}}. \tag{20}$$

Depending on whether the reagents and ISCs or only the ISCs participate in the sth elementary reaction we have:

$$w_s = k_s c_j, \tag{21}$$

$$w_s = k_s. \tag{22}$$

Let us construct the kinetic graph for given mechanism in such a way that for every elementary step from the mechanism (which may be reversible or nonreversible) there corresponds an edge. This kinetic graph is nonoriented. Comparing equations (1) and (17), and taking into account the isomorphism between a graph and a mechanism, it is evident that the number of independent reaction routes in a complex reaction system M_K (kinetic basis) with a linear mechanism given by the Horiuti Rule will be equal to the cyclomatic number of the corresponding nonoriented kinetic graph. This means that each independent route in the reaction mechanism corresponds to an independent cycle in the nonoriented kinetic graph. In the case when $M_K > M_T$, the difference $M_K - M_T$ or $\Omega - M_T$ will represent the number of stoichiometrically dependent or null routes. We use in this case a nonoriented kinetic graph because the Horiuti Rule gives the stoichiometric balance for elementary steps (but not for elementary reactions) participating in the reaction mechanism. For example, the cyclomatic number for the oriented graph in Fig. 1.1 is $\Omega_{or} = 6 - 3 + 1 = 4$, and for the nonoriented graph $\Omega_{nonor} = 3 - 3 + 1 = 1$. For the mechanism (4) only one stoichiometrically independent route exists and $M_T = \Omega_{nonor}$.

The Horiuti Rule for linear mechanisms is the chemical equivalent of the cyclomatic number equation, which expresses general topological characteristics of a graph (mechanism).

We now illustrate what has been said so far with an example of a complex multiroute reaction (the isomerization of butenes over Co-Mo/Al$_2$O$_3$ (22)) with a linear mechanism:

	I	II	III	IV
1) $1\text{-}C_4H_8 + Z = 1\text{-}C_4H_8Z$	1	1	0	0
2) $\text{cis-}2\text{-}C_4H_8 + Z = \text{cis-}2\text{-}C_4H_8Z$	-1	0	1	0
3) $\text{trans-}2\text{-}C_4H_8 + Z = \text{trans-}2\text{-}C_4H_8Z$	0	-1	-1	0
4) $1\text{-}C_4H_8Z = \text{cis-}2\text{-}C_4H_8Z$	1	0	0	1
5) $1\text{-}C_4H_8Z = \text{trans-}2\text{-}C_4H_8Z$	0	1	0	-1
6) $\text{cis-}2\text{-}C_4H_8Z = \text{trans-}2\text{-}C_4H_8Z$	0	0	1	1

I) $1\text{-}C_4H_8 = \text{cis-}2\text{-}C_4H_8$

II) $1\text{-}C_4H_8 = \text{trans-}2\text{-}C_4H_8$

III) $\text{cis-}2\text{-}C_4H_8 = \text{trans-}2\text{-}C_4H_8$

IV) $0 = 0$ (23)

Here Z is a free site on the catalyst surface, and

$1\text{-}C_4H_8Z$, $\text{cis-}C_4H_8Z$, and $\text{trans-}C_4H_8Z$ are ISCs for three butenes.

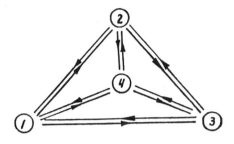

Figure 1.2

This mechanism corresponds to the reaction graph represented in Fig. 1.2.

For this particular mechanism the stoichiometric basis and the route basis are given by the following equations:

$$M_T = 3 - 1 = 2$$
$$M_K = 6 - 4 + 1 = 3.$$

The total number of cycles in the kinetic graph is given by:

$$\Omega = 6 - 4 + 1 = 3.$$

We have two stoichiometrically independent routes in the thermodynamic basis and three independent routes in the kinetic basis of this mechanism. It follows that one route of the kinetic basis is stoichiometrically dependent. If we assume that the third route is stoichiometrically dependent, two opportunities exist for making M_T equal to M_K:

a) Remove from the mechanism one elementary step so $M_K = 5 - 4 + 1 = 2$ and $\Omega = 5 - 4 + 1 = 2$

b) Change the stoichiometric numbers of the elementary steps participating in the third route so that it will become a null route. Such an opportunity is presented as route IV). The reaction along this route proceeds at a finite rate without influencing the stoichiometry of the process.

The relationship between the reaction rates for the reagents and the rates along the linearly independent stoichiometric routes may be expressed by the following equations:

$$r_1 = -R_1 - R_2$$
$$r_2 = R_1 \qquad (24)$$
$$r_3 = R_2$$

If we take into account the third stoichiometrically dependent route, system (24) becomes:

$$r_1 = -R_1 - R_2$$
$$r_2 = R_1 - R_3 \qquad (25)$$
$$r_3 = R_2 + R_3$$

where the reagents labeled 1 to 3 are, respectively, 1-butene, cis-2-butene, and trans-2-butene.

1.6 Deriving the Kinetic Equations by Graph Theory Methods

The application of graph theory methods for deriving kinetic equations of heterogeneous catalytic reactions is based upon the so-called Rule of Mason; this is also known in American literature as the Shannon–Mason Rule of Cycles. Although established by Shannon in 1941 (23), the rule acquired great popularity after its rediscovery by Mason in 1955 (24, 25). A strict proof for the validity of the Rule of Mason for multiroute linear mechanisms was presented only recently by Evstigneev and Yablonskii (26), where both an inductive proof and a proof based on the Rule of Krammer are set forth.

1.6.1 Equation for expressing the concentrations of ISCs
Application of the Rule of Mason in chemical kinetics was made for the first time in the work of Vol'kenshtein and Gol'dshtein (3, 4) and leads, by analogy to electrotechnics, to the following expression for the concentrations of the ISCs participating in the studied mechanism:

$$X_\alpha = \frac{C \cdot D_\alpha}{D}, \qquad (26)$$

where X_α is the concentration of the ISC α, C is the total concentration of the ISCs which is usually set equal to one, D_α is the base determinant of the kinetic graph for the vertex X_α, and D is the base determinant of the kinetic graph (mechanism) and is equal to sum of the base determinants of all the vertices in the kinetic graph, i.e.

$$D = \sum_{i=0}^{n} D_i, \qquad (27)$$

where n is the number of vertices in the kinetic graph.

We shall call the sum of the weights of all the maximum trees of the graph ending on a given vertex the base determinant of that vertex. The chemical meaning of the base determinant is very simple: it shows all the reaction pathways by which ISC α is produced from all other ISCs participating in the reaction mechanism. The base determinant of the mechanism contains pathways for producing all ISCs in the mechanism.

We now give an expressions for the rate of a catalytic reaction obtained from the Rule of Mason.

Let suppose that the reaction proceeds under a steady state condition and for every vertex we can write the equation:

$$X_i \sum_{i \neq j} w_{ij} = \sum_{i \neq j} X_j w_{ij}. \qquad (28)$$

This means that every kinetic graph is equivalent to a steady state system (11). Evaluation of the kinetic equation by the methods of graph theory is equivalent to solving system (11).

1.6.2 Equation of Vol'kenshtein and Gol'dshtein for the reaction rate of independent routes

Following Vol'kenshtein and Gol'dshtein (2, 3), the reaction rate along the stoichiometrically independent route r_p may be expressed by the equation:

$$r_p = r_p^+ - r_p^- = \frac{\sum\limits^{N} (w_p^+ - w_p^-) D_{pn}}{D_0}, \qquad (29)$$

where r_p^+ and r_p^- are the rates along the cycle, which correspond to a given reaction route, in the forward and reverse directions, w_p^+ and w_p^- are the weights of the cycle in the forward and reverse directions. They are equal to the product of the weights of the arcs in the cycle, N is the number of cycles remaining after the cycle p shrinks to a point in its base, D_{pn} are the base determinants of the cycles remaining after the cycle p shrinks to a point in its base, and D_0 is the base determinant of the graph with a root in the null vertex.

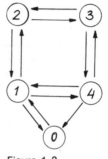

Figure 1.3

The graph represented in Fig. 1.3 has two independent reaction routes. The reaction rate for routes 0140 and 12341 is given by the equations:

$$r_{0140} = \frac{w_{0140}D_{0140} + w_{012340}\cdot 1}{D_0}$$

$$r_{12341} = \frac{(w_{12341}^+ - w_{12341}^-)D_{12341} + w_{012340}\cdot 1}{D_0} \tag{30}$$

where $D_{012340} = 1$ because cycle 012340 contains all vertices of the graph.

For noncatalytic reactions a null vertex is added to the kinetic graph, while for catalytic reactions in corresponding kinetic graphs no null vertex appears because the role of the null vertex is that of a free active site on the catalyst's surface.

Then, according to Vol'kenshtein and Gol'dshtein (2, 3), the rate of an enzyme reaction may be expressed in the form:

$$r = \frac{E \sum_j k_i D_i}{D},$$ (31)

where E is the total enzyme concentration, k_i are rate constants for product formation from vertex i of the graph, D_i is the base determinant of vertex i, i is the number of types of active centers in the enzyme, and D is the base determinants of the kinetic graph.

If the enzyme possesses only one type of active site, equation (31) transforms into:

$$r = \frac{E k_i D_i}{D}.$$ (32)

This form of the kinetic equation may also be used in the description of single-route chemical processes.

The following characteristic features of equation (29) are worthy of comment:

1. The analogy with electric circuits made earlier is certainly not an obvious one, and in the works of Vol'kenshtein and Gol'dshtein (2, 3), a requirement exists that the kinetic graph for noncatalytic reactions contain a null vertex with a concentration equal to 1. For catalytic reactions the role of the null vertex is that of a free catalytic surface.

2. The rate along the route is expressed by the rate along the whole cycle, taking into account all the elementary steps involved in the cycle, i.e. in all cases we solve Temkin's system of equations for the steady state (11).

1.6.3 Equation of Klibanov, Slinko, Spivak and Timoshenko for the Reaction Rate of an Independent Reaction Route

Another, more rational approach has been proposed by Klibanov, Slinko, Spivak and Timoshenko in (15). A route basis is selected in such a way that there is at least one step, called the basic step, in each stoichiometrically independent route that is not included in the other routes. Consider a reaction mechanism that consists of S elementary steps and has Q independent routes. The system (11) for this mechanism is thus divided into the two subsystems (33) and (34). Let us renumber the steps so that those involved in only one route receive numbers from 1 to Q.

Then we have:

$$r^{(q)} = r_q^+ - r_q^- \qquad q = 1, 2, \ldots, Q \tag{33}$$

and also:

$$\sum_{p=q+1}^{P} \sigma_t(p) R(p) = r_t^+ - r_t^- \qquad t = Q+1, Q+2, \ldots, S. \tag{34}$$

Hence, system (33) states in explicit form the rates along the independent routes as a function of the concentrations of the reagents and the ISCs.

The reaction rate of an elementary reaction is given by equation (7):

$$r_s = k_s X_\alpha \prod_{j=1}^{m} P_j^{\mu_j s}. \tag{7}$$

By substituting (7) and (26) into (33), we finally obtain an expression for the reaction rate along the qth route:

$$r^{(q)} = k(q)^+ X_\alpha \prod_{j=1}^{m} P_j^{\mu_j s} - k(q)^- X_\beta \prod_{j=1}^{m} P_j^{\mu_j s}. \tag{35}$$

According to equation (35), the rate along the qth route is not determined by all species participating in the cycle's q steps, but only by the reaction rate for the base elementary step s, which proceeds from a vertex designated α towards a vertex designated β. This implies that, instead of solving the system of equations for the steady state (11), we solve a considerably simpler system of steady state equations (33). This makes equation (35) very convenient for solving practical problems.

1.6.4 Simplified equations of Yablonskii and Bykov for mono and multiroute reactions

Simplified equations, proposed by Yablonskii and Bykov et al. (27–30) are derived using equation (35). These simplified equations are useful for the theoretical analysis of the mechanisms of catalytic reactions and the structure of kinetic equations.

1.6.4.1 *Equation for single route reaction*

Equation (35) may be rewritten in the following way (27):

$$r^{(q)} = \frac{w_s^+ \sum\limits_{k=1}^{N_i} W_{k,i} - w_s^- \sum\limits_{k=1}^{N_{i+1}} W_{k,i+1}}{\sum\limits_{i=1}^{n} \sum\limits_{k=1}^{n} W_{ki}} \tag{36}$$

where w_s^+ and w_s^- are the weights of the arcs in the forward and reverse directions for the step s, $W_{k,i}$ and $W_{k,i+1}$ are the weights of the tree for vertices i and $i + 1$, and N_i and N_{i+1} are the numbers of trees at the corresponding vertices

It then follows that:

$$D_i = \sum_{k=1}^{N_i} W_{k,i} \quad \text{and} \quad D_{i+1} = \sum_{k=1}^{N_{i+1}} W_{k,i+1} \tag{37}$$

are the base determinants of the vertices i and $i + 1$, while

$$D = \sum_{i=1}^{n} \sum_{k=1}^{n} W_{ki}, \tag{38}$$

is the sum of the base determinants for all n graph vertices.

The graph of a single-route complex catalytic reaction with all its steps reversible is represented in Fig. 1.4

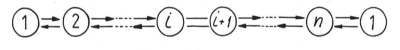

Figure 1.4

After the corresponding transformations, which we omit here, equation (36) may be rewritten as follows:

$$r = \frac{\overset{n}{\underset{i=1}{A}} w_i^+ - \overset{n}{\underset{i=1}{A}} w_i^-}{\sum\limits_{i=1}^{n} W_{\text{str},i} + \sum\limits_{i=1}^{n} W_{\text{rev},i} + \sum\limits_{i=1}^{n} \sum\limits_{k=1}^{n} W_{\text{mix},i}}, \tag{39}$$

where $\sum\limits_{i=1}^{n} W_{\text{str},i}$ is the sum of all the forward trees. We call those trees "forward" whose arcs traverse the graph in the forward direction. $\sum\limits_{i=1}^{n} W_{\text{rev},i}$ is the sum of all the reverse trees. We call those trees "reverse" whose arcs traverse the graph in the reverse direction. $\sum\limits_{i=1}^{n} \sum\limits_{k=1}^{n} W_{\text{mix},i}$ is the sum of all the mixed trees. Trees possessing arcs directed in both the forward and reverse directions are termed "mixed". n is the number vertices in the graph.

The total number of trees is n^2, which is the sum of n forward trees, n reverse trees and $n(n-2)$ mixed trees. This classification of the spanning tree gives a simple algorithm for calculating the base determinant for a one-route mechanism.

If step j of the mechanism is irreversible, we have $w_j^- = 0$ and

$$r = \frac{\prod\limits_{i=1}^{n} w_i^+}{\sum\limits_{i=1}^{n} W_{\text{str},i} + \prod\limits_{i \neq j}^{n} w_{\text{rev},i} + \sum\limits_{i=1}^{n} \sum\limits_{k=1}^{n} W_{\text{mix},i}}. \tag{40}$$

If there are two or more irreversible steps, we find that

$$r = \frac{\prod\limits_{i=1}^{n} w_i^+}{\sum\limits_{i=1}^{n} W_{\text{str},i} + \sum\limits_{i=1}^{n} \sum\limits_{k=1}^{n} W_{\text{mix},i}}, \tag{41}$$

and, when all the steps are irreversible, that

$$r = \frac{\prod\limits_{i=1}^{n} w_i^+}{\sum\limits_{i=1}^{n} W_{\text{str},i}}. \tag{42}$$

For a cycle with a pendant vertex connected to a vertex

(schematically represented in Fig. 1.5), we obtain the same equation with the minor difference that the weights of the kinetic graph arcs entering the vertex α are divided by the quantity $(1 - K_\alpha)$, where K_α is the ratio of the weights of the forward and reverse steps leading toward the pendant vertex.

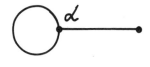

Figure 1.5

The numerator in equation (39) is referred to as the "cyclic characteristics". It represents the kinetic equation for the net reaction with all its steps obeying the Law of Mass Action, whereas the denominator reflects the rate decrease caused by the reagents and the products.

1.6.4.2 *Kinetic equation for a biroute reaction*

The mechanisms of biroute reactions can be divided into three groups (28) as indicated in Fig. 1.6. The kinetic equations along the two independent routes have the following form:

For the first type:

$$r^{(1)} = \frac{r_c^{(1)} D_\alpha^{(2)}}{D}$$

$$r^{(2)} = \frac{r_c^{(2)} D_\alpha^{(1)}}{D},$$

(43)

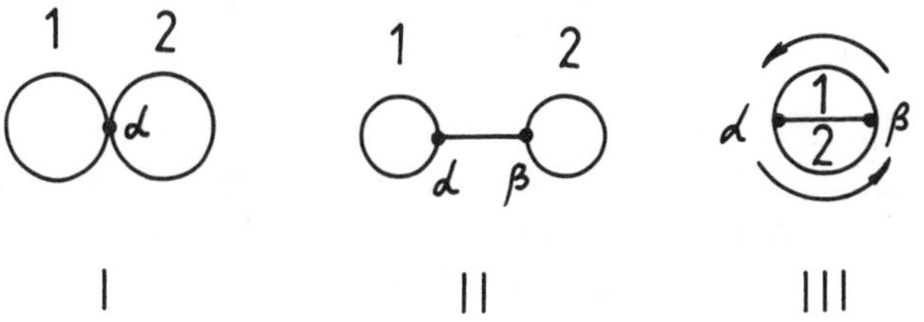

I   II   III

Figure 1.6

where $r^{(1)}$ and $r^{(2)}$ are the rates along the first and second cycles respectively, $r_c^{(1)}$ and $r_c^{(2)}$ are the cyclic characteristic of the two cycles. D is the sum of the base determinants of the graph, and $D_\alpha^{(2)}$ and $D_\alpha^{(1)}$ are the sum of the weights of the trees for vertex α in cycles 2 and 1 respectively.

For the second type:

$$r^{(1)} = \frac{r_c^{(1)} D_\beta^{(2)} w_{\alpha\beta}^-}{D}$$

$$r^{(2)} = \frac{r_c^{(2)} D_\alpha^{(1)} w_{\alpha\beta}^+}{D}$$

(44)

where $D_\alpha^{(1)}$ and $D_\beta^{(2)}$ are the sums of the weights of the trees for vertices α and β in cycles 1 and 2 respectively. $w_{\alpha\beta}^+$ and $w_{\alpha\beta}^-$ are the weights for the forward and reverse reactions for the step $\alpha = \beta$, e.g.

For the third type:

$$r^{(1)} = \frac{r_c^{(1)} D_2 + r^*}{D}$$

$$r^{(2)} = \frac{r_c^{(2)} D_1 - r^*}{D}$$

(45)

where D_1 and D_2 are the sums of the weights of the trees for vertices α and β in the corresponding cycles divided by the weight of the net reaction, $r_c^{(1)}$ and $r_c^{(2)}$ are the cyclic characteristics of the two cycles, r^* is referred to as the conjugation characteristics of the cycles and is given by:

$$r^* = \prod_{i=2}^{n_1} w_{i1}^+ \prod_{i=2}^{n_2} w_{i2}^- - \prod_{i=2}^{n_1} w_{i1}^- \prod_{i=2}^{n_2} w_{i2}^+$$

(46)

where $w_{i1}^\pm$ and $w_{i2}^\pm$ are the weights of the steps in the corresponding cycles, and n_1 and n_2 are the numbers of steps in each cycle. The common step in both cycles is denoted by the number 1.

It should be noted that this simplified form of biroute mechanism is not very useful for practical purposes because of difficulties in determinating the base determinants of the graph.

1.6.4.3 Kinetic equation of a complex multiroute reaction

The rate of an elementary step in a complex multiroute reaction is given by the following equation (29, 30):

$$r_s = \frac{\sum_{i=1}^{M} W_i r^*}{D} \qquad (47)$$

where M is the number of cycles in which the step s participates, W_i is the cyclic characteristics of the ith cycle expressed by the formula:

$$W_i = \prod_v w^+(v) - \prod_v w^-(v), \qquad (48)$$

where v is the number of arcs in the ith cycle,

$w^+(v)$ and $w^-(v)$ are the weights of all the arcs involved in the cycle in both positive and negative directions, and r^* is the conjugation parameter of the ith cycle. This assumes the form:

$$r^* = \sum_H w(H), \qquad (49)$$

where H is number of cycles of the kinetic graph which are left after cycle i shrinks to a point in its base.

The conjugation parameter is a quantity showing how the compounds not included in a given cycle affect its reaction rate. In terms of graph theory, this is the weight of the oriented forest having its roots at the vertices of a cycle in which the arcs of the cycle are not included. In other words, this is the base determinant of a given cycle.

The denominator in equation (47) characterizes the complexity of the reaction.

Although equation (47) has the same structure as equation (29) proposed by Vol'kenshtein and Gol'dshtein, its components have a different chemical meaning:

(a) Equation (29) gives the rate along an independent reaction route and uses as a basis all elementary steps included in the route. The further development of this approach led to equation (32), which uses a simplified basis of basic elementary steps for every cycle to express the rate along independent cycles.

(b) Equation (47) gives the rate of a individual elementary step of the reaction mechanism using as a basis set all cycles in which this particular step is involved.

(c) For noncatalytic reactions with linear mechanisms the kinetic equation (29) needs a kinetic graph which contains null vertices, while equations (35) and (47) can be applied without such a restriction.

1.7 Application of Graph Theory to Nonsteady State Processes

Vol'kenshtein and Gol'dshtein were the first to apply graph theory to investigation of the kinetics of nonsteady state processes (3, 31). We present here a short summary of their method.

In nonsteady state reactions, the concentration derivatives with respect to time are not equal to zero and, instead of stationary conditions (12), a system of differential equations is obtained:

$$\frac{dx(\tau)}{d\tau} + x(\tau) \sum_{r=1}^{n} a_{tr} = \sum_{s=1}^{n} a_{st} x(\tau) \quad t = 1, 2, \ldots, n-1. \tag{50}$$

System (50) is nonlinear and cannot be solved exactly if a_{tr} and a_{st} are functions of the time τ. If we assume that the concentrations of the reagents are much larger than those of the ISCs, then a_{st} and $a_{tr}s$ can be taken as independent of time.

Let the integral transformation of Laplace–Carson be applied to system (50). The functions of time $x(\tau)$ can then be substituted by the $x*(q)$ transformation functions, and the time derivatives by the $qx*(q) - qx(0)$ values, where q represents reciprocal time and the $x(0)$ are the initial concentrations of the intermediates. The Laplace–Carson procedure transforms the system of differential equations (50) into a system of algebric equations with respect to $x*(q)$ thus:

$$x_t.(q) \sum_{r=1}^{n} a_{tr}^* = \sum_{s=1}^{n} a_{st}^* x_s(q) \quad t = 1, 2, \ldots, n-1 \tag{51}$$

with

$$\sum_{t=1}^{n} x_t^*(q) = 1.$$ (52)

The coefficients $a*$ coincide with the coefficients a of the equations describing stead-state reactions for all steps, with the exception of the steps going to the initial state:

$$a_{ij}^* = \begin{bmatrix} a_{ij} & j = 1 \\ a_{i,j+q} & j = 1 \end{bmatrix}$$ (53)

This implies that for nonsteady state steps system (51) is equivalent to a graph.

The graph of the nonsteady state reactions is obtained from the graph of the stationary reaction by the addition of new arcs extending from each vertex to the vertex of the free catalytic surface. The weight of each arc is equal to q.

The graph obtained is solved by the method described above to give the Laplace–Carson transformation function of the reaction rate:

$$r^*(q) = \frac{\alpha_2 q^{n-2} + \alpha_3 q^{n-3} + \ldots + \alpha_{n-2}}{q^{n-1} + \beta q^{n-2} + \ldots + \beta_{n-1}},$$ (54)

where n is the number of vertices in the graph.

The quantity q refers to reciprocal time, making possible a comparison of the different processes in terms of the time necessary to reach the steady state. When $q \to 0$ the stationary value of r is obtained:

$$r^*(0) = r = \frac{\alpha_{n-2}}{\beta_{n-2}}.$$ (55)

Only the term with the highest power of q is left in the expression for $r^*(q)$ at the beginning of the process:

$$r^*(q \to \infty) = \frac{\alpha_2}{q},$$ (56)

or

$$r(\tau \to 0) = \alpha_2$$ (57)

Therefore, α_2 can be evaluated from the initial slope of the $r(\tau)$ curve. The exact form of the original is obtainable from the $r^*(q)$ transformation with the help of special tables or computer calculations.

Graph theory also allows us to explain oscillations in reacting systems. Intermediate compounds are formed just before the stationary state is reached with concentrations equal to the stationary concentrations. Additional edges appear in the non-steady state with weight equal to q. They connect each ISC with the vertex corresponding to the free surface of the catalyst. Two opposite connections of weight q appear if the graph has at least three vertices. A situation can then occur in which the decrease of the intermediates (at the expense of the opposite arc) could lead to an increase of the concentration of another intermediate, and hence oscillations might occur.

In his monograph, Clarke (32) makes extensive use of graph theory to study the stability of complex reaction mechanisms. Graph theory is also used to describe kinetics of chemical reactions complicated by diffusion of reagents into solid catalysts (33).

1.8 Evaluating Base Determinants of a Kinetic Graph

The application of the methods of graph theory to experimental data from kinetics studies of heterogeneous catalytic reactions,is strongly restricted by the difficulties in calculating the base determinants of the vertices in kinetic graphs. While for a simple graph it is possible to find the base determinants by hand, for a complex mechanism this is often a difficult task.

Several algorithms (34, 35) for solving this problem have been proposed, but they proved to be inefficient. Three algoriths (6, 39), proposed by the author for the efficient calculation of the base determinants of kinetic graphs, are now presented.

The base determinant of a vertex of an oriented kinetic graph is the sum of the weights of the spanning trees entering it. The spanning tree is a connected oriented subgraph of the initial graph containing no cycles. In order to transform a graph into a spanning tree it is necessary to remove an arc from each cycle i.e. a total of

Ω arcs (where Ω is the cyclomatic number of the graph) which form an aggregate called a "section". In this case we are interested in every elementary reaction in the mechanism. For this reason an oriented kinetic graph should be used; the calculated cyclomatic number of the graph is different and larger than the kinetic basis of the mechanism. The cyclomatic number of the oriented kinetic graph we denote as Ω_{or}. For a given graph containing Ω cycles, the maximal possible number of spanning trees T will be given by the formula:

$$T = \prod_{i=1}^{\Omega} t_i, \tag{58}$$

where t_i is the arc number in cycle i.

In fact, the spanning tree number will be smaller because there are always arcs that belong to more than one cycle. In (16) formulas are presented for the calculation of the spanning tree numbers of graphs having 2, 3 and 4 cycles (provided an arc does not belong to more than two cycles). The spanning tree number in a given graph may be estimated exactly using the Lantiery-Trent Theorem, which states that:

For the graph $G(X, P)$ on n vertices a square ($n \times n$) matrix S is formed in such a way that the diagonal element S_{ii} equals the number of arcs incidental to vertex X_i, while $S_{ij} (i \neq j)$ equals the negative number of arcs connecting vertices X_i and X_j of the graph. The maximum spanning tree number then equals the main minor of the matrix S:

$$T = q_s, \tag{59}$$

where q_s is the value of the main minor of the matrix S.

According to the Rule of Mason, the base determinant of a given vertex includes only spanning trees that end on this vertex. The number of needed spanning trees is actually smaller than that given by equation (59).

For greater clarity we illustrate our results by the model graph R, as shown in Fig. 1.7. The cyclomatic number of this graph is:

$$\Omega_{or} = 5 - 3 + 1 = 3 \tag{60}$$

Then we have the three independent cycles:

1) 121;   2) 232;   3) 3213.

or

1) 12. 21;   2) 23. 32;   3) 32. 21. 13.

$$(61)$$

The spanning tree number for the graph R using the Lantiery-Trent Theorem is obtained in the following way:

$$S_R = \begin{bmatrix} 3 & -2 & -1 \\ -2 & 4 & -2 \\ -1 & -2 & 3 \end{bmatrix}$$

$$(62)$$

$$q_R = \begin{bmatrix} 4 & -2 \\ -2 & 3 \end{bmatrix} = 4.\,3 - (-2).\,(-2) = 8$$

The maximum possible number of spanning trees of the graph is thus equal to 8.

1.8.1 Method for finding spanning trees based on the cyclomatic matrix

In order to find all the spanning trees we excise from the graph q_R times sections of Ω_{or} arcs combined from all the arcs of the graph.

To determine the set of sections, we use the cyclomatic matrix A_R of the graph R in which the rows correspond to the independent cycles, and the columns to the arcs of the graph. If an arc participates in a given cycle, the corresponding matrix element is equal to the arc; otherwise it is equal to zero. We obtain the following matrix for the graph R:

	arcs	12	21	23	32	13
	cycles					
	I	12	21	0	0	0
$A_R =$	II	0	0	23	32	0
	III	0	21	0	32	13

$$(63)$$

Taking one nonzero element from each cycle each time, we obtain the following 12 sections:

1) 12.32.21   4) 12.23.21   7) 21.32.21   10) 21.23.21

2) 12.32.32   5) 12.23.32   8) 21.32.32   11) 21.23.32   (64)

3) 12.32.13   6) 12.23.13   9) 21.32.13   12) 21.23.13

As shown above, we have eight spanning trees for the graph R and accordingly the sections should number 8. Therefore, there are sections in (64) that do not yeidl spanning trees. To eliminate them from further study we use the following two rules:

1. Rule One — If a section contains two identical arcs, it is set to zero as it does not lead to the formation of a tree. A cycle remains in the graph because the arc number in the section will be smaller than Ω_{or}.

2. Rule Two — If a certain section appears several times in the obtained section set, two possibilities exist: if it appears an even number of times the section is omitted; if it appears an odd number of times it is included only once.

If we apply these rules to sections (64), we find that the sections 2), 7), 8), and 10) are equal to zero (because they contain two identical arcs each). Finally, we obtain the following $q_R = 8$ sections:

$$
\begin{array}{ll}
1)\ 12.32.21 & 5)\ 12.23.13 \\
2)\ 12.32.13 & 6)\ 21.32.13 \\
3)\ 12.13.21 & 7)\ 21.23.32 \\
4)\ 12.23.32 & 8)\ 21.23.13
\end{array}
\qquad (65)
$$

To obtain the spanning trees, we form the vector $P = P$ $(12, 21, 23, 32, 13)$ composed of all the graph arcs. Consecutively taking away the sections (65) from the vector P yields the desired spanning trees:

$$
\begin{array}{ll}
1)\ 13.23 & 5)\ 32.21 \\
2)\ 21.23 & 6)\ 12.23 \\
3)\ 13.32 & 7)\ 12.13 \\
4)\ 21.13 & 8)\ 12.32
\end{array}
\qquad (66)
$$

For the final tree selection we must take into account the following rules:

3. Rule Three — If the spanning tree consists of more than one term, then all the terms must end on one and the same vertex. All

the trees that do not obey this rule are set equal to zero.

4. Rule Four — If the spanning tree consists of more than one term and some of its terms are part of other terms, then the first terms are set to zero.

If we apply the third and fourth rules to the trees (66), trees 2) and 7) are equal to zero as they end on different vertices.

We finish up with the following trees:

$$
\begin{array}{lll}
1) \; 13.23 & 3) \; 21.13 & 5) \; 12.23 \\
2) \; 13.32 & 4) \; 32.21 & 6) \; 12.32
\end{array} \tag{67}
$$

1.8.2 Method for finding spanning trees based on the section matrix

The spanning tree of a kinetic oriented graph on n vertices contains $n - 1$ arcs. As the graph is connected, all the spanning tree arcs should be connected to one another. The tree structure in this case is determined by the position of the $n - 1$ vertices and the nth vertex will be dependent. We construct the section matrix B so that the rows correspond to the independent graph vertices and the columns to the graph arcs. The B matrix elements will be equal to the arcs if a given vertex is incident to the corresponding arc; otherwise they will be equal to zero.

The graph spanning trees are obtained by taking each time one nonzero element from each row. As the graph is connected, we obtain the graph trees directly. For the graph R we find:

$$
B_R =
\begin{array}{c|ccccc}
\text{arcs} & 12 & 21 & 23 & 32 & 13 \\
\hline
\text{vertices} & & & & & \\
1 & 12 & 21 & 0 & 0 & 13 \\
3 & 0 & 0 & 23 & 32 & 13
\end{array} \tag{68}
$$

When we perform the operation (described above), the following trees are obtained.

$$
\begin{array}{lll}
1) \; 12.32 & 4) \; 21.32 & 7) \; 13.32 \\
2) \; 12.23 & 5) \; 21.23 & 8) \; 13.23 \\
3) \; 12.13 & 6) \; 21.13 & 9) \; 13.13
\end{array} \tag{69}
$$

Applying rules 1–4th the trees (69), we find that the tree 9) must equal zero because it consists of two identical arcs. The trees 3) and 5) end on different vertices and are also equal to zero. Finally, we therefore obtain the same result (67) as by the previous method.

1.8.3 Method for finding spanning trees using the transformation matrix G

Let the n vertices of the kinetic planar graph R be numbered with natural numbers. This graph corresponds to a square matrix G_R of dimension $(n \times n)$ whose elements are defined thus:

$$G_{ij}^{(0)} = \begin{bmatrix} j \text{ if there is an arc from vertex } i \text{ to vertex } j \\ 0 \text{ if there is no arc between } i \text{ and } j \text{ or } i = j \end{bmatrix} \quad (70)$$

We transform the matrix $G^{(0)}$ n times and the new elements are then defined by the relations:

$$G_{ij}^{(r)} = \begin{bmatrix} G_{ij}^{(r-1)} + G_{ir}^{(r-1)} \cdot G_{rj}^{(r-1)} & \text{if } i \neq r \text{ or } j \neq r \\ G_{ij}^{(r-1)} & \text{if } i = r \text{ or } j = r \end{bmatrix} \quad (71)$$

where $r = 1, 2, \ldots, n$.

Upon performing the transformation (71), we use the restriction rules given below:

1. $G_{ij}^{(r)} \cdot 0 = 0 \cdot G_{ij}^{(r)} = 0$

2. $G_{ij}^{(r)} \cdot G_{pq}^{(r)} \neq G_{pq}^{(r)} \cdot G_{ij}^{(r)}$

3. $G_{ij}^{(r)} + G_{pq}^{(r)} = G_{pq}^{(r)} + G_{ij}^{(r)}$

4. $G_{ij}^{(r)} \cdot (G_{pq}^{(r)} \cdot G_{kl}^{(r)}) = (G_{ij}^{(r)} \cdot G_{pq}^{(r)}) \cdot G_{kl}^{(r)} = G_{ij}^{(r)} \cdot G_{pq}^{(r)} \cdot G_{kl}^{(r)}$

5. $G_{ij}^{(r)} \cdot (G_{pq}^{(r)} + G_{kl}^{(r)}) = G_{ij}^{(r)} \cdot G_{pq}^{(r)} + G_{ij}^{(r)} \cdot G_{kl}^{(r)}$

6. $G_{ij}^{(r)} \cdot G_{kl}^{(r)} \ldots G_{mn}^{(r)} \ldots = 0$ if $m = q$ or $n = q$

(72)

for all elements of row q, excluding the qth column element.

7. $G_{ij}^{(r)} \cdot G_{pq}^{(r)} \ldots G_{kl}^{(r)} \cdot G_{ij}^{(r)} \ldots = 0$

After we perform the transformation (71) on matrix G, each matrix element on the left hand side is multiplied by the number of the row in which it appears and a new matrix H is thereby obtained. The matrix elements of matrix H have the following significance:

1. The diagonal elements, H_{ii}, equal the sum of the cycles passing through vertex i. This means that the diagonal element shows the number of routes in which the given ISC participates, and it gives the routes themselves. On the basis of the diagonal elements, the basic steps for each route may be readily determined.

2. All the nondiagonal elements, H_{ij}, equal the sum of the reaction pathways connecting intermediate i with intermediate j.

The matrix $G^{(0)}$ for the graph R has the form:

$$G^{(0)} = \begin{bmatrix} 0 & 2 & 3 \\ 1 & 0 & 3 \\ 0 & 2 & 0 \end{bmatrix} \tag{73}$$

After performing the transformations (71) and taking into account the constraints (72), we obtain for each step the following matrices:

$$G^{(1)} = \begin{bmatrix} 0 & 2 & 3 \\ 1 & 12 & 3+13 \\ 0 & 2 & 0 \end{bmatrix}$$

$$G^{(2)} = \begin{bmatrix} 21 & 2 & 3+23 \\ 1 & 12 & 3+13 \\ 21 & 2 & 213 \end{bmatrix}$$

$$G^{(3)} = \begin{bmatrix} 21+321 & 2+32 & 3+23 \\ 1 & 12+32+132 & 3+13 \\ 21 & 2 & 23+213 \end{bmatrix}. \tag{74}$$

We then multiply on the left each row's elements with its row number to obtain matrix H:

$$H = \begin{bmatrix} 121+1321 & 12+132 & 13+123 \\ 21 & 212+232+2132 & 23+213 \\ 321 & 32 & 323+3213 \end{bmatrix} \tag{75}$$

Taking consecutively each of the sections (65), we set to zero the vertices participating in it, ánd substitute the zero values into the elements of matrix H. Each nondiagonal matrix element containing a zero arc is set to zero; the remaining nonzero elements form a spanning tree. As the diagonal elements of H contain only cycles, they should not be taken into account in the above considerations. Rules 3 and 4 (given above) are applied to the spanning trees obtained and we thus finally determine the necessary set of spanning trees. For example, for section 1) 12. 32. 21 we obtain:

$$H = \begin{bmatrix} 0 & 0 & 13 \\ 0 & 0 & 23 \\ 0 & 0 & 0 \end{bmatrix}, \tag{76}$$

so the spanning tree will be 13. 23.

For section 2) 12. 32. 21 we find:

$$H = \begin{bmatrix} 0 & 0 & 0 \\ 21 & 0 & 23 \\ 0 & 0 & 0 \end{bmatrix}. \tag{77}$$

The tree 21. 23 obtained ends on two different vertices and for this reason is equal to zero.

The quickest and the most direct method is clearly the second one, i.e. that using the section matrix. The third method, although a little more complicated, gives greater insight. Moreover, as we shall see, it can be used to yield the cyclic characteristics, for base step selection, and in determining the conjugation parameter of a given cycle. This is why it is more convenient to use the first or second method in simple cases; for more complicated mechanisms, however, it is advisable to use the third method.

1.9 Determination of the Cyclic Characteristics of a Given Route

The cyclic characteristics are represented by the difference between the products of the weights of the steps participating in a given route in the forward and reverse directions. The most convenient method is to use the diagonal element, H_{ii}. It contains all the cycles in which a given vertex participates. Recording their weights in the forward and reverse directions in the sequence in which the arcs appear in the cycle, we obtain the cyclic characteristics of the given cycle.

1.10 Determination of the Conjugation Parameter of a Given Cycle

The conjugation parameter of a given cycle is the weight of the oriented forest, ending in that cycle though the cycle arcs are not included or it is the base determinant of the subgraph which is obtained when the cycle shrinks to a vertex. The conjugation parameter may be readily determined by the above methods as follows:

Methods based on cyclomatic and section matrices

After shrinking the given cycle to the vertex, the subgraph obtained is treated according to the proposed methods. For every cycle this operation should be performed separately.

Method based on the G matrix:

(1)   Fix the vertices of the cycle whose conjugation parameter is to be determined.

(2)   The diagonal elements of matrix H are deleted.

(3)   The H matrix rows corresponding to the cycle vertices are deleted.

(4)   Delete the H matrix columns corresponding to vertices not included in the cycle.

(5)   From the now reduced matrix terms delete the elements containing edges included in the cycle.

(6)   The remaining nonzero elements represent the desired conjugation parameter.

Using matrix (75) and the rules given above, we obtain for the graph R the following conjugation parameters for the independent cycles:

$$P_{121} = H_{31} + H_{32} = 321 + 32 = 32$$
$$P_{232} = H_{12} + H_{13} = 12 + 132 + 13 + 123 = 12 + 13 \qquad (78)$$
$$P_{3213} = 1$$

1.11 Application of graph-theoretical treatment to the experimental data

We consider now as an example the single-route reaction of nitrobenzene hydrogenation to aniline over industrial copper catalysts proceeding according to the equation:

$$C_6H_5NO_2 + 3H_2 = C_6H_5NH_2 + 2H_2O. \qquad (79)$$

Experimental data revealed that the reaction can be best described by the following mechanism (40):

1. $C_6H_5NO_2 + Z = C_6H_5NO_2Z$
2. $C_6H_5NO_2Z + H_2 \rightarrow C_6H_5NOZ + H_2O$
3. $C_6H_5NOZ + H_2 \rightarrow C_6H_5NHOHZ$ (80)
4. $C_6H_5NHOHZ + H_2 \rightarrow C_6H_5NH_2Z + H_2O$
5. $C_6H_5NH_2Z = C_6H_5NH_2 + Z$

This mechanism corresponds to the graph given in Fig. 8, where vertex 1 corresponds to Z a free site on the catalyst surface, and Vertices 2–5 correspond respectively to the ISCs $C_6H_5NO_2Z$, C_6H_5NOZ, C_6H_5NHOHZ, and $C_6H_5NH_2Z$.

The weights of the elementary reactions in steps 1–5 are correspondingly:

$$
\begin{aligned}
w_1^+ &= k_1^+ \cdot P_1 \qquad w_1^- = k_1^- \\
w_2^+ &= k_2^+ \cdot P_3 \\
w_3^+ &= k_3^+ \cdot P_3 \qquad\qquad\qquad (81) \\
w_4^+ &= k_4^+ \cdot P_3 \\
w_5^+ &= k_5^+ \qquad\quad w_5^- = k_5^- \cdot P_2
\end{aligned}
$$

The matrix $G^{(0)}$ assumes the following form:

$$G^{(0)} = \begin{bmatrix} 0 & 2 & 0 & 0 & 5 \\ 1 & 0 & 3 & 0 & 0 \\ 0 & 0 & 0 & 4 & 0 \\ 0 & 0 & 0 & 0 & 5 \\ 1 & 0 & 0 & 0 & 0 \end{bmatrix}. \tag{82}$$

After performing the transformations (71), and observing the rules (72), the matrix H is obtained:

$$H = \begin{bmatrix} 121+151+123451 & 12 & 123 & 1234 & 12345+15 \\ 21+23451 & 212+234512 & 23 & 234 & 2345 \\ 3451 & 34512 & 345123 & 34 & 345 \\ 451 & 4512 & 45123 & 451234 & 45 \\ 51 & 512 & 5123 & 51234 & 512345+515 \end{bmatrix} \tag{83}$$

The cyclic characteristics, C, of the equation are given by:

$$C = \prod_{s=1}^{5} w_s^+ = k_1^+ P_1 \cdot k_2^+ P_3 \cdot k_3^+ P_3 \cdot k_4^+ P_4 \cdot k_5^+ = \alpha_1 P_1 P_3^3 \tag{84}$$

where $\alpha_1 = k_1^+ k_2^+ k_3^+ k_4^+ k_5^+$.

Applying the methods described above, we obtain the following spanning trees with the corresponding weights:

	Trees	Weights
1.	34. 45. 51. 21	$k_1^- k_3^+ k_4^+ k_5^+ P_3^2$
2.	23. 34. 45. 51	$k_2^+ k_3^+ k_4^+ k_5^+ P_3^3$
3.	34. 45. 51. 12	$k_1^+ k_3^+ k_4^+ k_5^+ P_1 P_3^2$
4.	45. 51. 12. 23	$k_1^+ k_2^+ k_4^+ k_5^+ P_1 P_3^2$
5.	51. 12. 23. 34	$k_1^+ k_2^+ k_3^+ k_5^+ P_1 P_3^2$
6.	12. 23. 34. 45	$k_1^+ k_2^+ k_3^+ k_4^+ P_1 P_3^3$
7.	21. 15. 34. 45	$k_1^- k_3^+ k_4^+ k_5^- P_2 P_3^2$
8.	15. 23. 34. 45	$k_2^+ k_3^+ k_4^+ k_5^- P_2 P_3^3$

$$\tag{85}$$

The base determinants for the individual vertices are given by the following expressions:

$$D_1 = k_1^- k_3^+ k_4^+ k_5^+ P_3^2 + k_2^+ k_3^+ k_4^+ k_5^+ P_3^3$$
$$D_2 = k_1^+ k_3^+ k_4^+ k_5^+ P_1 P_3^2$$
$$D_3 = k_1^+ k_2^+ k_4^+ k_5^+ P_1 P_3^2 \tag{86}$$
$$D_4 = k_1^+ k_2^+ k_3^+ k_5^+ P_1 P_3^2$$
$$D_5 = k_1^+ k_2^+ k_3^+ k_4^+ P_1 P_3^3 + k_1^- k_3^+ k_4^+ k_5^- P_2 P_3^2 + k_2^+ k_3^+ k_4^+ k_5^- P_2 P_3^3$$

The base determinant, D, takes the form:

$$D = \alpha_2 P_3^2 + \alpha_3 P_3^3 + \alpha_4 P_1 P_3^2 + \alpha_5 P_2 P_3^2 + \alpha_6 P_1 P_3^3 + \alpha_7 P_2 P_3^3, \tag{87}$$

where $\alpha_2 = k_1^- k_3^+ k_4^+ k_5^+$
$$\alpha_3 = k_2^+ k_3^+ k_4^+ k_5^+$$
$$\alpha_4 = k_1^+ k_5^+ (k_3^+ k_4^+ + k_2^+ k_4^+ + k_2^+ k_3^+)$$
$$\alpha_5 = k_1^- k_5^- k_3^+ k_4^+$$
$$\alpha_6 = k_1^+ k_2^+ k_3^+ k_4^+$$
$$\alpha_7 = k_5^- k_2^+ k_3^+ k_4^+.$$

The reaction rate will be expressed by the following equation:

$$r = \frac{\prod_{s=1}^{5} w_s^+ \prod_{s=1}^{5} w_s^-}{D} = \frac{C}{D}. \tag{88}$$

which is obtained from equation (27). By substituting equations (84) and (87) into equation (88), we find that:

$$r = \frac{\alpha_1 P_1 P_3^3}{\alpha_2 P_3^2 + \alpha_3 P_3^3 + \alpha_4 P_1 P_3^2 + \alpha_5 P_2 P_3^2 + \alpha_6 P_1 P_3^3 + \alpha_7 P_2 P_3^3}. \tag{89}$$

Dividing equation (89) by $\alpha_2 P_2^2$ yields:

$$r = \frac{k_1 P_1 P_3}{1 + k_2 P_3 + k_3 P_1 + k_4 P_2 + k_5 P_2 P_3 + k_6 P_2 P_3}, \tag{90}$$

where $k_1 = \dfrac{\alpha_1}{\alpha_2};$

$$k_2 = \frac{\alpha_3}{\alpha_2};$$

$$k_3 = \frac{\alpha_4}{\alpha_2} \,;$$

$$k_4 = \frac{\alpha_5}{\alpha_2} \,.$$

$$k_5 = \frac{\alpha_6}{\alpha_2} \,; \text{ and}$$

$$k_6 = \frac{\alpha_7}{\alpha_2} \,;$$

Equations (84), (87) and (90) reveal the complex character of the kinetic constants obtained.

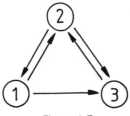

Figure 1.7

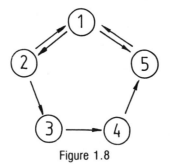

Figure 1.8

Let us associate the graph R in Fig. 1.7 with the mechanism of oxidation of nitrogen oxides by carbon monoxide over a silver catalyst (41). The reaction proceeds along two stoichiometrically independent routes:

	I	II
1) $NO + Z = ZNO$	1	0
2) $ZNO + NO \rightarrow N_2O + ZO$	1	0
3) $N_2O + Z \rightarrow N_2 + ZO$	0	1
4) $ZO + CO \rightarrow Z + CO_2$	1	1

$$(91)$$

I $2NO + CO = N_2O + CO_2$

II $N_2O + CO = N_2 + CO_2$

The graph R and the mechanism (91) become isomorphic if the vertices of the graph are denoted in a new way. We substitute vertex 2 by Z, vertex 1 by ZNO, and vertex 3 by ZO. The weights of the elementary reactions are then respectively:

$$w_1^+ = k_1^+ P_1 \qquad w_1^- = k_1^-$$
$$w_2^+ = k_2^+ P_1$$
$$w_3^+ = k_3^+ P_2$$
$$w_4^+ = k_4^+ P_3$$

$$(92)$$

where P_1, P_2, P_3 are the partial pressures of NO, N_2O and CO.

The base determinants are obtained from the spanning trees (67):

$$D_1 = D_{ZNO} = 32.\,21 = w_4^+ w_1^+ = k_1^+ k_4^+ P_1 P_3$$
$$D_2 = D_{ZN2O} = 12.\,32 + 13.\,32 = w_1^- w_4^+ + w_2^+ w_4^+$$
$$= k_1^- k_4^+ P_3 + k_2^+ k_4^+ P_1 P_3$$

$$(93)$$

$$D_3 = D_{ZO} = 12.\,23 + 21.\,13 + 13.\,32 = w_1^- w_3^+ + w_1^+ w_2^+ + w_2^+ w_3^+$$
$$= k_1^- k_3^+ P_2 + k_1^+ k_2^+ P_1^2 + k_2^+ k_3^+ P_1 P_2.$$

The denominator of the kinetic equation will be:

$$D = D_1 + D_2 + D_3 = k_1^+ k_2^+ P_1^2 + k_4^+(k_1^+ + k_2^+)P_1 P_3 + k_2^+ k_3^+ P_1 P_2$$
$$+ k_1^- k_4^+ P_3 + k_1^- k_3^+ P_2$$

$$(94)$$

To express the reaction rate we use equation (47):

$$r = \frac{\Sigma C_i r_i^*}{D}.$$

$$(95)$$

The numerator of the kinetic equation will consist of a cyclic characteristic and a conjugation parameter.

Let us determine the rate of step 4 for the above mechanism. This rate is denoted in the graph by arc 32. From the diagonal element H_{22} and H_{33} of matrix H (92), we see that arc 32 appears in cycles 323 and 3213. The cyclic characteristics will be respectively:

$$\begin{aligned}
\text{cycle 323} \quad & C_{323} = w_3^+ w_4^+ = k_3^+ k_4^+ P_2 P_3 \\
\text{cycle 3213} \quad & C_{3213} = w_1^+ w_2^+ w_4^+ = k_1^+ k_2^+ k_4^+ P_1^2 P_3.
\end{aligned} \tag{96}$$

The conjugation parameters of the two cycles are thus:

$$\begin{aligned}
\text{cycle 323} \quad & P_{323} = 12 + 13 = w_1^- + w_2^+ = k_1^- + k_2^+ P_1 \\
\text{cycle 3213} \quad & P_{3213} = 1
\end{aligned} \tag{97}$$

Finally, one obtains for the rate of the fourth elementary step of the mechanism the following expression:

$$\begin{aligned}
r_4 &= \frac{C_{3213} P_{3213} + C_{323} P_{323}}{D} \\
&= \frac{k_1^+ k_2^+ k_4^+ P_1^2 P_3 + k_3^+ k_4^+ P_2 P_3 (k_1^- + k_2^+ P_1)}{k_1^+ k_2^+ P_1^2 + k_4^+ (k_1^+ + k_2^+) P_1 P_3 + k_2^+ k_3^+ P_1 P_2 + k_1^- k_4^+ P_3 + k_1^- k_3^+ P_2}
\end{aligned} \tag{98}$$

We use for expressing the reaction rate on independent routes equation (35). Then for the reaction rates of routes I) and II) we have:

The basic elementary step for route I) is elementary step 2) and

$$r^{(1)} = \frac{D_1 P_1}{D} \tag{99}$$

The basic elementary step for route II) is elementary step 3) and

$$r^{(2)} = \frac{D_2 P_2}{D} \tag{100}$$

Finally, we have the following equations:

$$r^{(1)} = \frac{k_1^+ k_4^+ P_1^2 P_3}{k_1^+ k_2^+ P_1^2 + k_4^+ (k_1^+ + k_2^+) P_1 P_3 + k_2^+ k_3^+ P_1 P_2 + k_1^- k_4^+ P_3 + k_1^- k_3^+ P_2} \tag{101}$$

The basic elementary step for route II) is elementary step 3) and

$$r^{(2)} = \frac{k_1^- k_4^+ P_2 P_3 + k_2^+ k_4^+ P_1 P_2 P_3}{k_1^+ k_2^+ P_1^2 + k_4^+ (k_1^+ + k_2^+) P_1 P_3 + k_2^+ k_3^+ P_1 P_2 + k_1^- k_4^+ P_3 + k_1^- k_3^+ P_2}$$

(102)

The methods described here for deducing kinetic equations are very convenient for computer processing. This can be done manually when the graphs are of small dimensions.

1.12 Analysis of the Structure of the Kinetic Equations

The simplified forms of the kinetic equations obtained are very convenient for carrying out analyses on the structure of the kinetic models and on the relations between their parameters. Such analyses have been published by Yablonski, Bykov *et al.* (36–38) For this reason we state the final formulas without proof.

1.12.1 Relation Between the Rates of the Forward and Reverse Reactions

Equation (35) can be rewritten in the form:

$$r = \frac{k^+ \prod_{j=1}^{M} C_j^{n_j} - k^- \prod_{j=1}^{M} C_j^{m_j}}{D} = \frac{r^+ - r^-}{D},$$

(103)

where n_j and m_j are the overall orders of the forward and reverse reactions, C_j is the concentration of the reagents, k^+ and k^- are the rate constants for the forward and reverse reactions, M is the number of elementary steps involved in the reaction mechanism, and D is the sum of the basic determinants for all vertexes in the reaction graph.

Since not more than one molecule of the reagents participates in every elementary reactions, n_j and m_j represent the number of steps in which the jth compound participates. In the general case,

$$m_j = \epsilon m_j^*$$
$$n_j = \epsilon n_j^* \qquad j = 1, 2, 3, \ldots, N_A \tag{104}$$

where n_j^* and m_j^* are the stoichiometric coefficients of the net equation with minimal integer coefficients, ϵ is multiplicity of the reaction, and N_A is the number of the reagents.

Equation (103) can then be rewritten as:

$$r = r^+ - r^- = r^+ [1 - f(c)/K_p'] \tag{105}$$

where K_p is the equilibrium constant of the net reaction with minimal coefficients, and $f(c)$ is a function of the concentrations expressed thus:

$$f(c) = \prod_{i=1}^{M} c_j^{m_j^* - n_j^*} \tag{106}$$

The molecularity of the reaction is given by the following expression:

$$M_j(c) = \frac{\partial \ln r^+}{\partial \ln c_j} - \frac{\partial \ln r^-}{\partial \ln c_j} = \frac{\partial \ln \left[k^+ \prod_{i=1}^{M} c_j^{n_j} \right]}{\partial \ln c_j} - \frac{\partial \ln \left[k^+ \prod_{i=1}^{M} c_j^{n_j} \right]}{\partial \ln c_j} \tag{107}$$

Obviously,

$$M_j(c) = n_j^* \tag{108}$$

or, in other words, the molecularity of the reaction is equal to the stoichiometric coefficient of the jth compound in the net equation. The molecularity will also equal the total number of molecules of a given compound participating in all steps of the complex reaction.

If the mechanism consists only of irreversible steps, in some of which the jth reactant participates, then:

$$\frac{\partial \ln r}{\partial \ln c_j} \sum X_{ij}, \tag{109}$$

where X_{ij} is the concentration of the ith ISC that reacts with the jth initial reagent.

The molecularity can be also defined as:

$$M_j(T) = (E^- - E^+)/\Delta H_{pj}, \qquad (110)$$

where ΔH_{pj} is the heat effect corresponding to the net equation with the stoichiometric coefficient of the jth compound equal to 1.

The final form is obtained as:

$$M_j(T) = M_j(c) = n_j. \qquad (111)$$

The relation

$$\frac{r^+}{r^-} = \frac{f(c)^+}{f(c)^-} K_p(T) \qquad (112)$$

is not always valid for multiroute reactions. Here are some of the cases when it is valid:

(a) The cycles have only common vertices. In this case each step participates in only one simple cycle and there is no more than one representation of the cyclic characteristics.

(b) The step participates in several cycles, and all cycles but one are stationary and their cyclic characteristics are equal to zero.

(c) The step participates in simple cycles having the same net equations and proportional cyclic characteristics.

12.2 Dependence of the Reaction Rate on the Temperature

Equation (47) concerning the rate of the qth route of a multiroute reaction may be written as follows:

$$r^{(q)} = \frac{F_1^{(q)}(c, T)}{F_2^q(c, T)} = r^{(q)+} - r^{(q)-}, \qquad (113)$$

where c is the concentration of the observed substances, T is the temperature, $F_1(c, T)$ and $F_2(c, T)$ are functions of temperature and the concentrations, and q is the number of reaction routes taken into consideration.

If we introduce the function

$$x = \exp(-1/RT) \qquad (114)$$

and carry out the corresponding mathematical operations, we obtain for the rates of the forward and reverse reactions:

$$r^{+(q)} \approx x^n/(qx^p + \ldots + 1) \qquad n, q > 0, p > 1$$

$$r^{-(q)} \approx x^m/(qx^p + \ldots + 1) \qquad m, q > 0, p > 1,$$

(115)

where $(qx^p + \ldots + 1)$ is a polynomial having positive coefficients.

The qualitative behaviour of the whole system is determined by the relations between the parameters m, n, p. The functions $r^{+(q)}(x)$ and $r^{-(q)}(x)$ are either monotonic or pass through a single maximum. The $r^{(q)}(T)$ function is similar to that for $r^{(q)}(x)$.

(a) Exothermic Reactions

In this case $E^- > E^+$ and $m > n$. When x is small, the forward reaction proceeds, but when x is large the reverse reaction might proceed. Several cases are possible for intermediate values of x.

Let the activation energies of all steps in the forward reaction be smaller than those in the reverse reaction. In this instance of strongly exothermic reactions, $m > p$, e.g. $r^{(q)}(x)$ is a monotonically increasing function. $r_+^{(q)}(x)$ may also be a monotonically increasing function, or perhaps pass through a maximum. The rate of the net reaction has a single maximum.

Whenever some of the steps of the exothermic reaction are endothermic (hidden endothermicity), $n < m < p$. Then both functions have maxima and the rate of the net reaction can have two extremal points and, under certain conditions, thermal oscillations may occur in the rate of the reaction.

(b) Endothermic Reactions

In this case $E^+ > E^-$, e.g. $n > m$ and the same functional dependence is valid as for the exothermic reactions. Endothermic reactions with a hidden exothermic step occur most often in adsorption processes.

(c) Irreversible Reactions

When all the elementary steps are irreversible, the rate of the net reaction increases monotonically. If only certain of the steps are irreversible, the net rate may either increase monotonically or pass through a maximum, as it symptotically approaches zero.

1.12.3 Dependence of Reaction Rate on Pressure

It is convenient to write the equation for the rate of a catalytic reaction (35) in the following form:

$$r = \frac{k^+ \prod_{j=1}^{M} P_j^{m_j} - k^- \prod_{j=1}^{M} P_j^{n_j}}{D} = \frac{r^+ - r^-}{D}, \tag{116}$$

where the P_i are the partial pressures on the reagents.

The denominator in equation (116) is a polynomial containing the partial pressures having positive coefficients k_{ij}. From the structure of the kinetic equation, the maximum power of P_i in the denominator will be equal to the power in the numerator. Thus, the qualitative description of the reaction rate in terms of any partial pressure (all other partial pressures and the temperature being held constant) assumes the form:

$$r(p) \approx \frac{(x^q - a_0)}{x^q + q_1 x^{q-1} + \ldots + aq} \tag{117}$$

The following inequalities are then valid:

(i) $dr(p)/dp_i > 0$ for the starting compounds

(ii) $dr(p)/dp_i < 0$ for the products of the reaction. $\qquad$ (118)

Therefore, the rate for each variable increases monotonically. Upon simultaneous changing of two partial pressures of the starting compounds, the reaction rate passes through a maximum, while under the same conditions the reaction rate for the products of the reaction passes through a minimum.

The rate of a single-route reaction as a function of the partial pressures is characterized by a simple monotonic function. The latter cannot describe extremal or autocatalytic behaviour, which is characteristized by the reverse relations to those in (118), namely:

(iii) $dr(p)/dp_i < 0$ for the starting compounds

(iv) $dr(p)/dp_i > 0$ for the products of the reaction. $\qquad$ (119)

The dependence of the reaction rate on the net pressure also does not enable us to posit substantially new conclusions.

1.12.4 Observed Reaction Orders and Mechanism Characteristics

Each base determinant of a mechanism might include several steps in which one and the same reagent is consumed. Its weight will then be characterized by a power that includes the concentration of this reagent. This power shows the total number of consumed molecules of a given type. The denominator in equation (32) can be represented as a polynomial of the concentrations of a given reagent. If we assume that compound A reacts in p steps of the reaction mechanism, we have:

$$r = \frac{k[A]^p - W^-}{W_0 + W_1[A] + \ldots + W_p[A]^p},\tag{120}$$

where W_i and $W^- = \Pi_i w_i$ are functions of temperature and composition, and $[A]$ is the concentration of the reagent A.

$$K = \frac{\prod_i w_1^+}{[A]^p}\tag{121}$$

The observed reaction order for compound A is given by the formula:

$$m_A = \frac{\partial \ln r}{\partial \ln[A]}\tag{122}$$

If (120) is substituted into (122) and the differentiation is carried out, one obtains the final result:

$$m_A = \frac{\partial \ln r}{\partial \ln[A]} = \sum x_i + \frac{\sum_i D^-_{(i-1)}}{\sum_x D_x} + \frac{pr^-}{r},\tag{123}$$

where r and r^- are the rates of the net and reverse reactions respectively, Σx_i is the sum of the ISCs that react with A, and $\Sigma D^-(i-1)$ is the sum of the weights of the trees containing the reverse step of the x_ith compound.

The physical meaning of the three terms in equation (123) is as follows:

1. The first term is the sum of the stationary coverages of the ISCs reacting with the substance.
2. The second term is the part of the surface corresponding to the reverse reaction for the steps in which the substance participates.
3. The third term characterizes the reversibility of the reaction.

    If the reaction is irreversible, we know from equation (123) that:

$$m_A = \sum x_i , \qquad (124)$$

and that the observed order of the reaction can be used to evaluate the concentration of the surface compound reacting with A. In this case the value of the reaction order cannot be greater than one. The reaction order of reversible reactions cannot be greater than P, where P is the number of steps in which A participates.

1.12.5 Observable Energy of Activation

The observable energy of activation of a single-route reversible reaction can be expressed in the following way:

$$E_{obs} = \frac{\partial \ln r}{\partial(-1/Rt)} = \sum x_i E_i^+ + \frac{\sum\limits_{i=2}^{n+1} \Delta E_i^- D_{(i-1)}}{\sum\limits_{x} D_x} - \frac{\Delta H_p r^-}{r} , \quad (125)$$

where E_i^+ and E_i^- are the energies of activation for the forward and reverse reactions for step i, $\Delta H_p = \Sigma E_i^- - \Sigma E_i^+ = \Sigma_i \Delta H_{pi}$ is the heat effect for the net reaction, and $\Delta E_i = E_i^+ - E_{i-1}^-$ is the difference between the energies of activation of the two steps in which one and the same ISC participates.

The three terms in equation (125) respectively determine the effect of the ISC, the reverse steps, and the reversibility of the reaction. If one of the steps is irreversible, the last term in (125) is equal to zero. When all the steps are irreversible, we have:

$$E_{obs} = \sum_i x_i E_i^+ , \qquad (126)$$

and E_{obs} cannot be larger than the energy of activation of the separate steps. In the case of several reversible steps, E_{obs} can be larger than the maximum energy of activation of certain steps, but cannot be greater than ΣE_i. Finally, if all steps are reversible, E_{obs} can acquire an arbitrarily high value.

1.13 References

1. B. King, C. Altman, J. Phys. Chem., **60**, 1375, 1956.
2. M.I. Temkin, Dokl. Akad. Nauk USSR, 163, 615, 1965.
3. M.V. Vol'kenstein, B.N. Gol'dstein, Dokl. Akad. Nauk USSR, 170, 963, 1965.
4. M.V. Vol'kenstein, Physics of Enzymes, Sciences, Moscow, 1967.
5. K.V. Yatsimirskii, Intern. Chem. Eng., 15, 7, 1977.
6. L.A. Petrov, in Theoretical Problems of Kinetics of Catalytic Reactions, Ed. S.L. Kiperman, Moscow, 1983, p. 39.
7. O. Ore, Theory of Graphs, Amer. Math. Soc. Colloq. Publ., Vol. 38, 1962.
8. C. Berge, Theorie des Graphes et ses Applications, Dunod, Paris, 1958.
9. M. Swamy, K. Tholasiraman, Graphs, Networks and Algorithms, John Wiley, 1984.
10. P.H. Sellers, Proc. Natl. Acad. Sci., US 55, 693, 1966.
11. P.H. Sellers, SIAM J. Appl. Math. 15, 13, 1967.
12. P.H. Sellers, Arch. Rat. Math. Anal., 44, 23, 1977.
13. O.A. Hougen, R.M. Watson, Ind. Eng. Chem., 35, 529, 1943.
14. K.H. Yang, O.A. Hougen, Chem. Eng. Progr., 46, No. 3, 146, 1950.
15. M.V. Klibanov, M.G. Slinko, S.I. Spivac, V.I. Timoshenko, Upravliaemye systemi, Novosibirsk, No. 7, 64, 1970.
16. D. Bonchev, O.N. Temkin, D. Kamenski, React. Kinet. Catal. Lett., 15, 119, 1980.
17. D. Bonchev, D. Kamenski, O.N. Temkin, J. Comput. Chem., 3, 95, 1982.
18. D. Shopov, L. Petrov, Comp, Rend. Acad. Bul. Sci., 21, 253, 1968.

19. L. Petrov, A. Eliyas, D. Shopov, Applied Catalysis, 18, 87, 1985.
20. M.I. Temkin, Dokl. Akad. Nauk USSR, 152, 156, 1963.
21. M.I. Temkin, Theoretical Foundation of Selection and Catalyst Manufacturing, Novosibirsk, 1964.
22. A. Eliyas, L. Petrov, C. Vladov, D. Shopov, Appl. Catalysis, 33, 295, 1987.
23. Shannon, C.E., The Theory and Design of Linear Differential Equation Machines OSRD Rept 411, Jan. 1942 under the auspices of Sec D-2 (Fire Control) of the National Defense Research Committee.
24. S.J. Mason, "Feedback Theory: Further Properties of Signal Flow Graphs Proceedings IRE", Vol. 44, No. 7, 1956, pp 920–926, and MXT res. Lab. Electron. Tech. Rept., July 20, 1955.
25. C.S. Lorens, Flow Graphs for the Modelling and Analysis of Linear Systems, McGraw-Hill, Monographs in Modern Engineering Sciences, 1964.
26. B.A. Evstigneev, G.S. Yablonskii, Kinetics and Catalysis, 20, 1549, 1979.
27. G.S. Yablonskii, V.I. Bykov, Kinetics and Catalysis, 18, 1561, 1977.
28. G.S. Yablonskii, V.I. Bykov, Dokl. Akad. Nauk USSR, 238, 645, 1978.
29. V.A. Evstigneev, G.S. Yablonskii, V.I. Bykov, Dokl. Akad. Nauk, 245, 871, 1979.
30. G.S. Yablonskii, V.A. Evstigneev, A.S. Noskov, V.I. Bykov, Kinetics and Catalysis, 22, 738, 1981.
31. M.V. Volkenshtein, B.N. Gol'dshtein, Molec. Biol., 1, 52, 1967 (Russ.)
32. B.L. Clarke, Stability of Complex Reaction Networks, in Adv. Chem. Phys., 43, 1, 1980, edit. I. Prigogine and S.A. Rice.
33. D.M. Auslander, G.F. Osten, A. Perelson, G. Kliford, On Systems with Coupled Chemical Reactions and Diffusions, Bond Graph Modeling for Engineering Systems, eds. D. Karnopp and R. Rosenberg, New York, USA.
34. R. Bot, J. Mayberry, Matrices and trees. Economic activity analysis, New York, J. Wiley, 1954.
35. H.I. Fromm, Biochem. Biophys. Res. Commun., 40, 692, 1968.

36. G.S. Yablonskii, V.I. Bykov, Theor. and Exper. Chem., 15, 41, 1979.
37. G.S. Yablonskii, M.Z. Lazman, V.I. Bykov, React. Kinet. Catal. Lett., 20, No. 1-2, 73, 1982.
38. A.S. Noskov, G.S. Yablonskii, Kinetics and Catalysis, 23, 191, 1983.
39. L.A. Petrov, D.M. Shopov, React. Kinet. Catal. Lett., 7, 273, 1977.
40. L.A. Petrov, N.V. Kirkov, D.M. Shopov, Kinetics and Catalysis, 21, 1275, 1980.
41. N.E. Bogdanchicova et al., Kinetics and Catalysis, 21, 1275, 1980.

Chapter 2

CLASSIFICATION AND CODING OF CHEMICAL REACTION MECHANISMS

O.N. Temkin[1] and D. Bonchev[2]

[1] Lomonosov Institute of Fine Chemical Technology,
Moscow, USSR

[2] Higher School of Chemical Technology, 8010 Burgas, Bulgaria

Introduction

Studies on the mechanisms of catalytic and non-catalytic reactions undertaken over the past 15–20 years have led to significant progress in the theory of reaction mechanisms. Most of the reactions involving homogeneous, metal-complex, and enzymatic catalyses were shown to be no less complex in terms of their mechanism compared with the mechanisms of radical chain processes. Infact, they appear to be much more complicated. Numerous examples of complicated mechanisms can be found in the literature[1-7]. At present, multiroute mechanisms (with 2 to 4 reaction routes), involving as many as 8 intermediates and up to 12 elementary steps, are widely known to exist even in heterogeneous catalysis by metals and nonmetals[8-10] where the simplest two-step schemes have hitherto been very popular. The existence of many routes and elementary steps is the most important general feature of the mechanisms of catalytic and also many noncatalytic reactions.

Contemporary chemical kinetics and the theory of reaction mechanisms are characterized not only by increased complexity of the mechanisms (hypotheses of mechanisms) but also by the considerable number of hypotheses (the possible mechanisms describing each reaction). Cases are known where the mechanism of formation of a certain product in a complicated multiroute mechanism incorporates completely different sequences of elementary steps and intermediates[11,12] even in the case of reactions that have one linearly independent stoichiometric equation[12]. The greater mechanistic complexity and high number of hypotheses raise the issue of the formalization and automation of the procedure adopted for the generation of hypotheses.

The importance of the problem may be assessed from the example of conjugated catalytic reactions in which the acrylic, propionic, succinic and maleic acid esters are prepared from CO and C_2H_2 in alcoholic solutions of $Pd(I)$ complexes[11,13]. The total number of hypotheses, generated on the simplest possible assumption that the final product is formed in a single sequence of intermediates, is 1344!

The first general theory for generating the maximal number of hypotheses was proposed by Sellers[14,15]; it was based on an elegant group-theoretical formalism. Other approaches to the problem are also known[8,11,16-19]. To develop an automated procedure for generating all possible hypotheses, however, a convenient coding of the mechanisms is needed to create a database of mechanisms.

The mechanism of each complicated reaction contains two kinds of information: chemical (or physicochemical) and topological (structural). The chemical information is governed by the nature, and composition of the intermediates and transition states, as well as by the reactivity of intermediates. Topological information expresses the mechanistic structure (in the intermediate space). It is determined by the number of reaction routes and intermediates, as well as by the differing ways in which the routes are connected.[20] Information that can be abstracted from kinetic data is very useful in the primary selection and discrimination of hypotheses. The best manner of expressing structural information is by means of the cyclic graphs proposed by Temkin[21] (hereinafter these graphs will be called "kinetic graphs").

In the case of mechanisms whose elementary steps incorporate one intermediate to the left and right of the reaction equality (called by Temkin "linear mechanisms"), each edge in the cyclic graph stands for an elementary step of the reaction mechanism, i.e. for a pair of mutually reversed elementary reactions. Each vertex of the kinetic graph corresponds to a certain intermediate while the linearly independent reaction routes are represented by graph cycles. For example, the mechanism of the water vapour methane conversion over *Ni* incorporates two independent routes, five intermediates, and six steps; it is depicted by kinetic graph 1.

1

Figure 2.1

The mechanism itself is given by the following scheme:

1. $CH_4 + Z \rightleftarrows ZCH_2 + H_2$
2. $ZCH_2 + H_2O \rightleftarrows ZCHOH + H_2$
3. $ZCHOH \rightleftarrows ZCO + H_2$
4. $ZCO \rightleftarrows Z + CO$
5. $Z + H_2O \rightleftarrows ZO + H_2$
6. $ZO + CO \rightleftarrows Z + CO_2$

$$CH_4 + H_2O = CO + 3H_2$$
$$CO + H_2O = CO_2 + H_2$$

The following intermediates are included: Z is a reaction site on the nickel surface and is assumed to be bivalent; CH_2, $CHOH$, CO, and O are hemosorped radicals. In the same listing order these intermediates correspond respectively to vertices $1, 3, 4, 5$, and 2 in the kinetic graph.

Directions of graph edges corresponding to forward and reverse elementary reactions are not given in the graph because they coincide with those elementary step directions assumed to be positive in the mechanism. If the rate of a reverse elementary reaction (r_{-s}) is zero, such an edge becomes irreversible and uniquely oriented (an oriented edge = arc) and is denoted by an arrow in the graph. In using such depiction, one arrives at a directed graph (digraph) in which all the arc directions are given by the direction of the full circumference; however, only the irreversible elementary steps are assigned (graph 2).

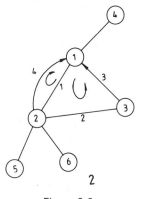

Figure 2.2

All products of the intermediate interactions 1, 2, and 3 with the reagents (vertices 4, 5, and 6) not participating in the elementary steps are depicted by pendant vertices (vertices of degree 1).

Non-linear elementary steps can also be depicted by cyclic graphs when additional, "secondary" edges are used.[22] The non-linear mechanisms can also be represented by so-called Vol'pert graphs.[23]

The use of cyclic graphs proved to be very fruitful for the deduction of kinetic equations and in the analysis of kinetic data for linear mechanisms. It should also be mentioned that Temkin's kinetic graphs provide a unique approach to both catalytic and non-catalytic reactions. In the latter case, one of the graph vertices contains a zero intermediate of concentration 1 not included in the kinetic equations. Thus, the use of kinetic graphs for non-catalytic reactions is justified only for mechanisms with at least one intermediate.

The structure of a kinetic graph expresses the structural information implicit in the reaction mechanism. In order to elucidate the connection between a graph's structure and topological features and formal kinetic laws, a mechanistic classification is required that should reflect these structural features of kinetic graphs. The most natural approach to such a classification is via graph theory[24-30], which has been applied with a great success to determining the complexity and coding of chemical compounds. The first attempts to use graphs in the characterization of the most typical and relatively simple mechanisms of heterogeneous reactions were reported in ref. 8. A classification has also been proposed for the two-route mechanisms of conjugated reactions proceeding from conjugation nodes.[31] This classification is, however, based on a rather heterogeneous set of criteria and not on the formal kinetic characteristics of mechanisms (graphs). The simplest types of two-route mechanisms and the relation between their structure and kinetics have been treated by Yablonskii and Bykov.[32]

We have developed general principles for the classification and coding of linear mechanisms on the topological basis of cyclic graphs.[33-36] This graph-theoretical approach unites the classification and coding with the procedure for deriving kinetic equations, and allows us to specify a hierarchy of mechanisms according to their degree of complexity.[37,38] Accordingly, the proposed approach could be very efficient in the creation of automated systems for chemical kinetics studies.

In this chapter, we outline the principles of classification, coding and decoding, and estimating the complexity of reaction mechanisms, as well as develop some approaches to identification of the topological structure of linear mechanisms.

2.1 Basic Classification and Coding Principles of Kinetic Graphs

Following on from the Introduction, the problem associated with the classification of linear mechanisms, when taking into account only structural information, is reduced to the classification of kinetic graphs. Such a classification, which provides a unique coding of the graphs, must fulfill certain specific requirements:

(i) The classification of the kinetic graphs (KG) must be complete, i.e. it must be capable of including all possible KGs, as well as offering tool for their generation. For these reasons, the classification should take into account all essential KG features. On the other hand, there exists an infinite number of cyclic graphs; and kinetics do not limit the number of reaction routes. The available methods, however, provide information on a limited number of routes, most frequently on 1 to 5 routes, though examples are known for 6 to 10-route mechanisms. Consequently, it seems sufficient to develop classification principles for linear mechanisms involving an acceptable practical number of routes (cycles) $M = 1\text{-}5$. Such a classification can readily be extended later on to larger M values.

(ii) The classification should reflect the KGs' basic topology, i.e. it should account for all essential factors including the number of cycles, their mutual location, the cycle linkages, the nature of these linkages, etc., but not the number of vertices in each cycle.

Our experience with KG classification has revealed that requirements (i) and (ii) can be fulfilled by making use of the concept of the so-called *supergraph*.[34] Each supergraph vertex represents a cycle in the initial KG while a pair of the supergraph vertices is connected by an edge when the respective KG cycles are directly joined (by means of a common edge(s), vertex(es), or bridge, but not through another cycle). The supergraph structure reflects the mechanism type (S) whereas the character of the supergraph edge (the way the cycles are connected) allows KG classes to be specified. For bicyclic graphs, where only one supergraph is possible ($\bigcirc\!-\!\bigcirc$), there are only three ways of linking the two cycles: a bridge, a common vertex or a common edge(s). For this reason, there are only three classes: A, B, and C.

a bridge a common vertex a common edge

class A   class B   class C

Figure 2.3

Further details on the structure of kinetic graphs are not of importance for the classification, though they are needed for their coding. We have solved the main problems for the classification and coding, those for the canonical depiction of KGs, and the canonical numbering of their vertices and edges.[34] The latter has been widely discussed in the literature.[39-45] In the case of kinetic graphs where a limited number of cycles and high symmetry are present, we prefer another approach[37] based on the generalized graph center concept.[46-49] The usefulness of this approach for graph cycle numbering was proven in the development of a convenient nomenclature and code for all (cata-, peri-, and corona-) fused benzenoid hydrocarbons[50], as well as in enunciating the principles of a universal nomenclature comprising all chemical compounds[51] (cf. Dr Goodson's Chapter in Volume 1 of this series). In our case, the numbering of the KG cycles reduces to the numbering of the supergraph vertices. The supergraph center determination, the procedure for the canonical numbering of its vertices by means of standard lattices, as well as the numbering of the KG vertices and edges, are discussed in some detail elsewhere.[37]

2.2 Classification and Coding of Undirected Kinetic Graphs

The proposed classification of linear mechanisms is based on a hierarchical system of criteria that allows an appropriate linear code to be constructed. Consider first the system of criteria, as well as the code for the undirected kinetic graphs (KG) without any pendant vertices:

1. Number of reaction routes (KG cycles), $M = 1, 2, 3, \ldots$
2. Number of intermediates (KG vertices), $N = 2, 3, 4, \ldots$
3. Mechanism type (the supergraph structure), $S = 0, 1, 2, \ldots$
4. Mechanism class is defined by the way in which two cycles in a KG are connected (or by the nature of the respective supergraph edge): bridging (class A), common vertex (class B), and common edge (class C).
5. Number of elements common to two cycles (the bridge length and the number of common edges: subclasses I and L).
6. Mutual position of two cycles linked to a third one (subclasses K).

In what follows we discuss these criteria and present the main results. The number of vertices in a KG, N, is not important for the classification, but it is convenient to introduce this criterion into the coding procedure immediately after the notation for the number of routes. The number of KG edges, E (the mechanism's elementary steps), is not regarded as a criterion because it is determined uniquely by the Horiuti rule[22]: $M = E - J$, where J is the number of linearly independent intermediates. In the case of non-catalytic reactions the number N includes the vertex with the so-called zero reagent.

2.2.1 Types of kinetic graphs (Supergraphs)

As mentioned in Section I of this chapter, the most essential KG characteristic is its supergraph which reflects the mutual connectivity of the cycles. For KGs with $M = 1$ and $M = 2$ there is only one supergraph and its type number, $S = 0$, is omitted for brevity. The type (serial) numbers of the supergraphs having $M = 1, 2, 3$, and 4 are presented in Table 2.1 together with the numbering of the KG cycles (the supergraph vertices).

Table 2.1. Supergraphs of kinetic graphs having one to four cycles. Their cycle numbering and serial numbers

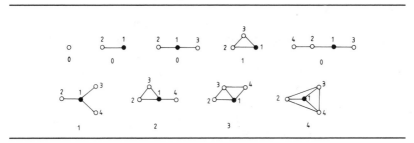

All supergraphs of the kinetic graphs having $M = 5$ are given in Table 2.2.

Table 2.2. Supergraphs of the five-route mechanisms. Their cycle numbering and serial numbers

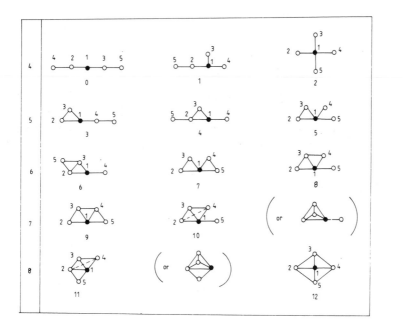

2.2.2 Classes and Subclasses of Kinetic Graphs

In specifying the classes of mechanisms related to the nature of the supergraph edge, one should take into account for KGs with $M \geqslant 3$ that the bridge between the two cycles must not contain edges or vertices common to a third cycle. In the general case, the notation of each KG class is given by the term $\Pi_{ij=1}^{M}X_{ij}$, where $X_{ij} = A, B$ or $C, i < j$, and $X_{ij} \neq 0$ when i, j are linked cycles. In this way, the class notation becomes $A^{x}B^{y}C^{z}$, where $x + y + z = E$, and E' is the number of supergraph edges. It is, however, preferable to avoid uniting all equal X_{ij} terms by means of the exponents x, y, and z, except in cases when two consecutive X_{ij} terms are the same, e.g. $BCC = BC^{2}$. When different depictions of the kinetic graph produce different class notations, the minimal code condition must be used, maintaining the priority order $A < B < C$.

Example:

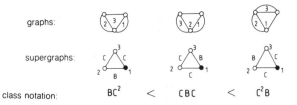

Figure 2.4

Proceeding from the linkage type of two KG cycles, another kind of class, B_{v}, is defined when the two cycles are connected by two or more disjoint vertices. This class corresponds to the introduction of a cycle (or a multiple edge) connecting two supergraph vertices.

Example:

Figure 2.5

In Tables 3 and 4 we present all the classes for $M = 1, 2, 3$, and 4. Their number is equal to $1, 3, 14$, and 99, respectively. Each class is illustrated by the corresponding representative graph with the smallest N that preserves its basic topology. More details concerning the procedure for KG generation will be reported elsewhere.[52]

The subclasses determined by the number of common edges forming a bridge between two cycles, as well as the edges common to two cycles, are denoted by A_I and C_L, respectively. No such subclasses for class B are possible because two common vertices form an edge common to two cycles (i.e. class C is thus obtained). For this reason, the graph containing the B_2 code constituent is regarded as belonging to an individual class but not to a subclass. The A_I and C_L subclasses have the same supergraph structure (the same basic topology) as the A and C classes whereas the B_V classes for $V = 1$ and $V = 2$ have a different topology (a different number of cycles).

Table 2.3. Classes of kinetic graphs having one, two, and three cycles and their notation

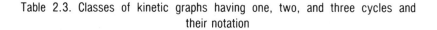

CYC-LES	TYPE SUPERGRAPH	CLASSES
1	o	⊝
2	o——● 2 1	A B C
3	3 – 0 o——●——o 2 1 3	A^2 B^2 C^2 AB AC BC
	3 – 1 ⟨³ 2 1	A^3 B^3 C^3 A^2B A^2C B^2C BC^2 B_2C^2

Table 2.4. Classes of kinetic graphs having four cycles and their notation

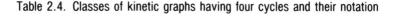

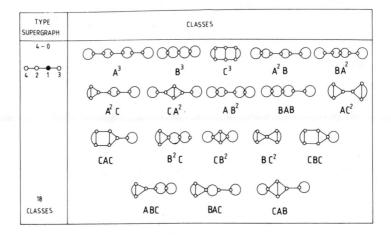

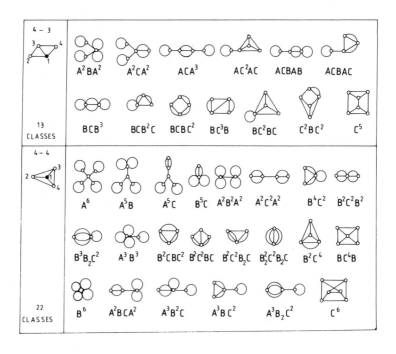

Table 2.4. *Continued*

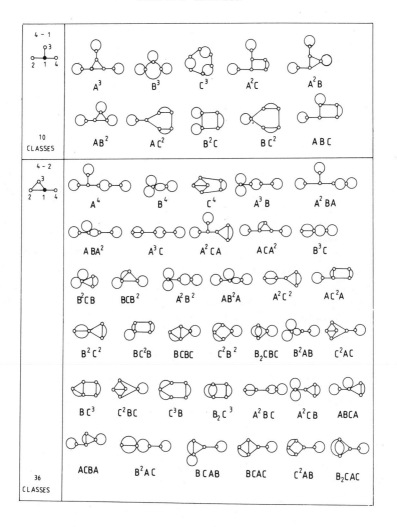

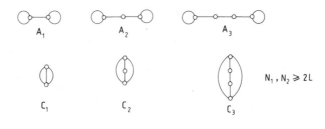

Figure 2.6

The mutual position of two cycles linked to a third one is reflected by a second subscript $K(A_{I,K}, B_{V,K}, \text{and } C_{L,K})$. The latter, more specifically, denotes the number of edges between the elements of two cycles common to a third one. This is illustrated by examples below where the second subscript $K = 0, 1,$ and 2 stands respectively for a distance of 0, 1, and 2 edges between the third cycle and the joint element between the first two cycles. The first subscript $I = 1$, $V = 1$, and $L = 1$ for the A, B, and C classes stands respectively for subclasses in which cycles 1 and 3 are connected by a bridge of unit length, a common vertex, and a common edge.

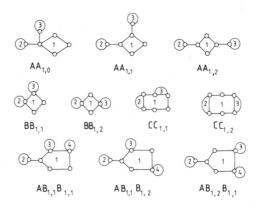

Figure 2.7

2.2.3 Linear Code: Applications

All the notations used in introducing the classification criteria can be united in a unique linear code:

$$M - N - S - A^x_{I,K} B^y_{V,K} C^z_{L,K} - N_1, N_2, \ldots, N_M$$

The code contains consecutively the number of cycles (routes) and vertices (intermediates) N, the supergraph type S, the classes (A, B, C, and B_2), the subclasses (the subscripts I, L, and K), and the number of vertices in each cycle N_i. (The full code actually includes a more detailed class notation which lists the connections between the pairs of cycles in increasing order, as seen from Table 2.4.) The code provides a one-to-one correspondence with the undirected kinetic graph described above. Some examples now follow for mechanisms described by undirected graphs.

The decomposition of formic acid in metals[53], as well as the reduction of carbon monoxide on silver[54]:

Figure 2.8

The conversion of methane by water vapor on nickel[55]:

Figure 2.9

The amination of alcohols and carbonyl-containing compounds in the presence of a molten iron catalyst[56]:

Figure 2.10

2.3 Coding of Graphs with Directed Edges or Pendant Vertices

The code for directed graphs is an extension of the undirected graph code. There are three possibilities for each KG edge: clockwise or counterclockwise orientation with respect to certain graph vertices, or both orientations (a reversible reaction step) which, however, are not marked on the kinetic graph. The symbols i, $\bar{i}$, and e will be used for these three cases respectively. The graph edges E are numbered according to increasing cycle number and moving clockwise in each cycle, proceeding from the first vertex common to the lowest number neighboring cycle. Thus, for instance, the numbering in the two examples given below proceeds respectively from the common vertex and the common edge in the bicyclic KG under examination.

In such a manner the sequence of all edge orientations, $E_1, E_2, \ldots, E_F$, is added to the undirected graph code. Each arc E_k is denoted by one of the three symbols specified above. To make the code unique, the priority order $i < \bar{i} < e$ should be used in conjunction with the minimal code criterion in the case of isomorphic directed graphs.

Example:

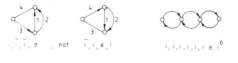

Figure 2.11

The last example also demonstrates how to shorten the code notation for arc directions:

$$i, i = i^2; \bar{i}, \bar{i} = \bar{i}^2; e, e = e^2$$

To illustrate the coding procedure, the directed graphs and their codes are given for the reaction mechanisms of a number of catalytic processes.

The four most important mechanisms in homogeneous catalytic

redox processes[57] are: (I) alternate oxidation and reduction of the catalyst; (II) activation by complex formation; (III) complex formation prior to the oxidation reaction (or reduction reaction); and (IV) complex formation prior to both stages of the redox conversion of the catalyst.

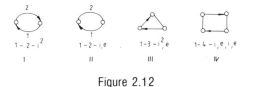

1 – 2 – ı²   1 – 2 – ı,e   1 – 3 – ı²,e   1 – 4 – ı,e,ı,e

    I   II   III   IV

Figure 2.12

Syntheses of vinylacetylene and 2-chlorovinylacetylene on the catalyst system $CuCI - CuCI_2 - NH_4CI - H_2O$[58]:

V   2 – 3 – C – 2,3 – ı, ı⁻²,e

Figure 2.13

Catalytic oxidation of CO to CO_2 in $PdCI_2$ solutions[12]:

VI   3 – 5 – 1 – B²C – 3,2,3 – ı⁴, ı⁻,e, ı

Figure 2.14

Oxidative chlorination of acetylene[59]:

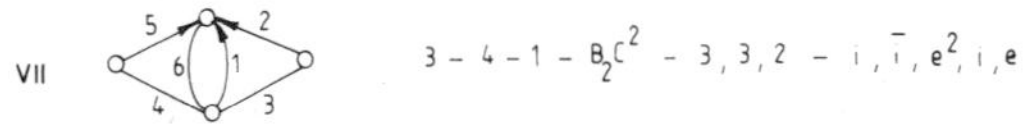

VII   3 – 4 – 1 – B₂C² – 3,3,2 – ı, ı⁻,e²ı,e

Figure 2.15

Graphs VIII and IX correspond respectively to the process of $\alpha - \beta$ butylene isomerization (chain mechanism)[60], and vinylchloride synthesis on $HgCI_2/C^{61}$. Their mechanisms are described by the same undirected graph, though they differ in the arc directions.

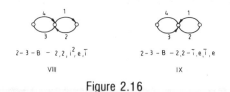

VIII   IX

Figure 2.16

Many catalytic processes (enzymatic and acid-base catalysis, catalysis using metal complexes, metals, oxides, etc.) are characterized by kinetic graphs with pendant vertices (vertices of degree one) joined to some vertex from the cyclic part of the graph (a base vertex). Such graphs are somewhat more complex than the corresponding cyclic graphs but their topology remains unchanged. The KG code in these cases is supplemented by the total number of pendant vertices, N_p, and, in increasing order, the number n_i of the base vertices to which the pendant vertex(es) is joined:

$$M - N - S - A^x_{I,K} B^y_{V,K} C^z_{L,K} - N_1, N_2, \ldots, N_M - E_1, E_2, \ldots, E_E - N_p : n_1, n_2, \ldots, n_p$$

Again, the minimal code criterion should be used when two or more isomorphic graphs exist; this occurs when the pendant vertex is joined to one of the several equivalent base vertices.

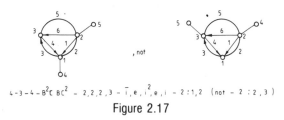

Figure 2.17

If several pendant vertices, p_i, are joined to the same base vertex n_i, the number p_i will appear in the code in brackets after the base vertex number:

$$- N_p : n_1(p_1), n_2(p_2), \ldots, n_p(p_p)$$

In this part of the code all the $p_i = 1$ are omitted for brevity.

The detailed algorithm for coding and decoding kinetic graphs for linear reaction mechanisms is given elsewhere.[34]

2.4 Approaches to the Classification and Coding of Nonlinear mechanisms

2.4.1 Types of Nonlinear Mechanism. Generalization of the Classification and Code for Linear Mechanisms

Nonlinear mechanisms are very common in heterogeneous catalytic reactions. They are also characteristic of chain reactions and, perhaps, of homogeneous catalysis involving metal complexes. Because of this, the classification of these mechanisms is of considerable interest.

Cyclic graphs with additional edges provide the possibility of conveniently depicting nonlinear mechanisms.[22, 9] Let us consider in detail the known types of nonlinearity and the ways of depicting them uniquely.

Taking into account the low probability for three-molecular elementary reactions, particularly in the case of collision of three intermediates, the types of nonlinearity can be reduced to two types of elementary bimolecular steps (with respect to the intermediates):

$$x_i \rightleftarrows x_j + x_k \tag{1-2}$$

$$x_i + x_j \rightleftarrows x_k + x_m \tag{2-2}$$

These types of nonlinearity are realized in the following specific kinds of elementary steps:

$$
\begin{aligned}
&(1-2)_a \ j = k &&(x_i \rightleftarrows 2x_j) \\
&(1-2)_b \ j \neq k && \\
&(2-2)_a \ i = j, \ k = m &&(2x_i \rightleftarrows 2x_k) \\
&(2-2)_b \ i = j, \ k \neq m &&(2x_i \rightleftarrows x_k + x_m) \\
&(2-2)_c \ i \neq j, \ k \neq m &&
\end{aligned}
$$

It is convenient to so depict the additional (broken-line) edge of the graph connecting vertices taking part in steps of type $(1-2)$ and $(2-2)$ as to distinguish between the cases of formation two equal or two different particles in the nonlinear steps:

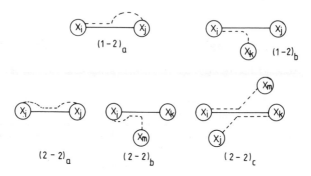

Figure 2.18

Vertex X_i in graphs incorporating nonlinearity of types 1, 2 and 5 can be a zero vertex.

Some examples of nonlinear mechanisms, depicted by means of additional edges, are presented below.

The mechanism of an unbranched chain reaction:

$$C_2H_6 = C_2H_4 + H_2$$

with elementary steps of thermal chain initiation and chain quadratic termination (graph 3):

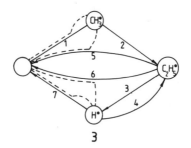

3

Figure 2.19

The branched chain reaction of hydrogen oxidation (simplified mechanism)[62] (graph 4):

$$\text{(1)} \quad H_2 + O_2 + M \rightarrow \underline{HO_2} + H + M*$$
$$\text{(2)} \quad H{\cdot} + O_2 \rightarrow OH + 0$$
$$\text{(3)} \quad OH + H_2 \rightarrow H_2O + H$$
$$\text{(4)} \quad H_2 + O \rightarrow OH + H$$
$$\text{(5)} \quad H + O_2 + M \rightarrow \underline{HO_2} + M*$$

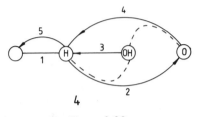

Figure 2.20

The catalytic synthesis of ammonia[9(p. 175)], (graph 5):

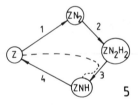

Figure 2.21

The Langmuir-Hinshelwood mechanism of heterogeneous catalytic processes (graph 6):

$$\text{(1)} \quad A + Z \rightleftarrows AZ$$
$$\text{(2)} \quad B + Z \rightleftarrows BZ$$
$$\text{(3)} \quad AZ + BZ \rightleftarrows PZ_2$$
$$\text{(4)} \quad PZ_2 \rightleftarrows P + 2Z$$

$$\overline{A + B = P}$$

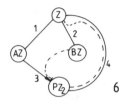

Figure 2.22

Depicting nonlinear mechanisms by means of cyclic graphs with additional edges can be used as a device to classify and code these mechanisms. As seen from the foregoing examples, the basic graph topology remains unaltered with the appearance of a non-linear elementary step and an additional edge; the numbers of cycles, vertices and edges stay constant. For these reasons, the classification and code we have developed for linear mechanisms could easily be extended to non-linear mechanisms by including additional information concerning the number and type of the nonlinear elementary steps, as well as the numbering of the inter-mediates included in each such step. We have already demonstrated that the nonlinear step can connect two type $(1 - 2)_a$ and $(2 - 2)_a$ intermediates, three type $(1 - 2)_b$ and $(2 - 2)_b$ intermediates, and four $(2 - 2)_c$ type intermediates.

2.4.2 Separate coding of linear and nonlinear subgraphs of a kinetic graph

Another natural approach to the classification and coding of non-linear mechanisms is a method based on the decomposition of the mechanism graph into two subgraphs, G_1 and G_2. G_1 is the subgraph of the linear elementary steps while G_2 is the subgraph of the nonlinear steps.

Subgraph G_1 is coded on the basis of criteria developed for cyclic graphs, supplemented by those of acyclic graphs. Subgraphs G_2 which are simple, mainly acyclic graphs (connected, as well as disconnected ones) can also be coded. Each vertex in G_2 is numbered by the labels of the vertices taking part in the nonlinear step. For example, graph 3 can be represented by means of sub-graphs G_{31} and G_{32}.

Linear subgraph G_{31}:

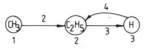

Figure 2.23

Nonlinear subgraph G_{32}:

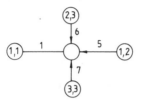

Figure 2.24

The vertex subscripts for G_{32} reflect the role of two inter-mediates in each of the elementary steps 6, 5, 7, and 1, indicating the numbers of these compounds as well as the formation of two intermediates in step 1.

Graph 6 has the following linear and nonlinear subgraph:

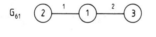

Figure 2.25

As a whole, the approach based on the decomposition of kinetic graphs into linear and nonlinear subgraphs provides for a complete description of chemical reactions, and at the same time reflects the type and stoichiometry of the nonlinear elementary steps.

2.5 Complexity of Mechanisms. Quantitative Estimation

In formulating hypotheses for the mechanism of a given complex reaction, and in using different procedures for the selection of one out of many hypotheses (discrimination of hypotheses), the question arises as to the hierarchy of the hypotheses. The intuitive principle of simplicity cannot play the role of a tool for the selection of hypotheses in the case of multiroute reactions because the number of vertices and cycles and the ways of linking cycles in the kinetic graph are already variables. Proceeding from linear mechanisms, we examine here possible approaches to the construction of a quantitative scale for mechanistic complexity or to the selection of a "complexity index".

2.5.1 Complexity of kinetic graphs

The first level of complexity of a linear mechanism is obviously related to the complexity of kinetic graph. The simplest graph characteristic that is related to kinetics is the degree of mutual connectedness of the cycles (for a constant number of cycles). This can be expressed quantitatively in terms of the total number of edges in the supergraph because each edge in the latter corresponds to a pair of connected cycles. Inspection of Table 2.1 indicates that in the case of three-route mechanisms this quantity discriminates between the two types of supergraph $3 - 0$ and $3 - 1$. In dealing with four-route mechanisms, a different numerical index is thus obtained for four out of five types of supergraph. It increases with the increasing supergraph serial number disregarding only the difference between the two acyclic supergraphs $4 - 0$ and $4 - 1$. In the case of the twelve five-route mechanisms (Table 2.2), this index has a weaker discriminating capacity, for it divides them into five groups of three or two supergraphs.

In order to discriminate the supergraph types and, furthermore, the classes of kinetic graphs, one can apply a supplementary numerical index which is closely related to the graph structure. Such quantities, called *topological indices*,[63-67, 29] are widely used in the quantitative characterization of graphs. They may also be regarded as a first approximation in evaluating the degree of graph complexity. The Wiener index,[68, 69] representing the sum of the distances in a molecular graph, is a very convenient measure for graph branching and cyclicity and, to a certain extent, for graph complexity. Calculations show that the Wiener index distinguishes among all the five types of supergraph having four cycles (Table 2.1) with the respective numerical values 10, 9, 8, 7, and 6 for the serial numbers 0, 1, 2, 3, and 4. The reciprocal value of the Wiener index, averaged over a unit distance, increases with increasing supergraph serial number: it equals 0.6, 0.667, 0.75, 0.857, and 1, respectively. The maximum relative complexity 1 thereby obtained corresponds to the complete graph K_4 (supergraph type $4 - 4$). For the supergraphs of the five-route mechanisms, however, a degeneracy appears that is typical for topological indices, i.e. equal values of the Wiener index are obtained for two non-isomorphic supergraphs. For example, the supergraphs with serial numbers 7 and 8 have the same Wiener index $W = 14$.

In chemical graph theory, ways to avoid the degeneracy are usually sought by constructing more sophisticated topological indices such as the Balaban index[70] or the Bertz index.[71] A combined topological index which reflects the topological structure of the graph as fully as possible has also been proposed.[72] The first conceptual models of molecular complexity were recently put forward[73-76], and some criteria for complexity measures have been formulated. The ultimate unique (and relatively simple) solution to the problem of estimating graph complexity is, however, still lacking.

2.5.2 Complexity of Kinetic Models

The second complexity level of chemical reaction mechanisms is the complexity level of the kinetic model corresponding to a given mechanism (or KG). Starting from the fact that ultimately the mechanism complexity will manifest itself in kinetics, it seems natural to look for a complexity index that reflects the graph complexity demonstrated in the kinetic model. Two kinds of kinetic models may be used for this purpose: (a) fractional-rational equations of the rate of routes in stationary or quasistationary processes having linear mechanisms; (b) systems of differential equations describing any kind of mechanism.

We have proposed the complexity index, K, based on the fractional-rational form of the rate laws for reaction routes[38]. This index is defined as the total number of weights (rate constants) for the elementary steps included in the numerator and denominator of the kinetic laws for all routes of a multiroute reaction. In calculating K it is convenient to use the Vol'kenstein-Gol'dstein algorithm[77,78] which is applicable to the derivation of the rate laws for the routes of all catalytic and noncatalytic reactions having linear mechanisms.

The rate for route p in a reaction having M routes is:

$$r_p = [x_i] \sum_{k=0}^{k \leqslant M-1} (\vec{C}_{pk} - \overleftarrow{C}_{pk}) D_{pk}/D_i, \tag{1}$$

where D_i is the vertex determinant (the sum of the weights of all trees containing this vertex), and C_{pk} is the length of the cycles corresponding to route p ($k = 0$) or the length of the cycle encompassing the cycle with $k = 0$. The cycle length is the product of the elementary step weights constituting the kth cycle; D_{pk} is the base determinant of the subgraph formed after contracting the pk cycle into a vertex (also called the algebraic complement of cycle pk). $[X_i]$ is the reagent concentration in vertex i (if the substance is a catalyst, active site or a zero reagent. In the latter case then $[X_0] = 1$).

This algorithm is very convenient, particularly for catalytic reactions, since if written as eqn. (2) it accounts for the material balance with respect to the catalyst:

$$r_p = [K]_\Sigma \frac{\displaystyle\sum_{k=0}^{k \leqslant M-1} (\vec{C}_{pk} - \overleftarrow{C}_{pk}) D_{pk}}{\displaystyle\sum_{i=1}^{N} D_i}. \tag{2}$$

Here $[K]_\Sigma$ is the total concentration of all the catalyst forms. In the case of surface reactions, $[K]_\Sigma = 1$.

The complexity index K is determined by the equation:

$$K = M(N - 1) T_p^d + N \sum_{p=1}^{M} T_p^n \tag{3}$$

where T_p^n and T_p^d are the total numbers of the spanning trees in the numerator and denominator for the kinetic laws governing r_p.

In this way the calculation of K is reduced to the calculation of the spanning trees of the kinetic graph (KG), as well as of some specific subgraphs. Formulas have been proposed[37] for calculating the number of spanning trees in KGs with $M = 2, 3$ and 4. Detailed calculations have also been made for the K index of 500 nondirected kinetic graphs referring to different classes of catalytic reaction mechanism.[38] Examples are presented in Table 5.

The influence of the different kinetic code constituents on the complexity index C for the reaction mechanism was proven[38] to be the following:

1. For a constant number of intermediates N, K increases with increasing number of routes M.

2. For a constant number of routes M and for a constant kinetic graph type and class, K increases with increasing number of intermediates N.

3. For a constant number of routes $(M = 3)$ and intermediates, changing the mechanism of types 3–0 to $3 - 1$ leaves K unaltered in some of the cases $(A^2 \to A^3,\ A^2 \to A^2B,\ AB \to A^2B,\ B^2 \to B^3,\ AC \to A^2C,\ BC \to B^2C,\ C^2 \to BC^2)$ whereas in the other cases K increases $(AB \to B^3,\ AC \to B^2C,\ A^2 \to A^2C,\ B^2 \to B^2C,\ BC \to BC^2,\ C^2 \to C^3,\ C^2 \to B_2C^2,\ AC \to B_2C^2,\ BC \to B_2C^2)$.

4. For a constant number of routes and intermediates and a constant type of reaction mechanism, the complexity index increases when going from a certain class to another one with larger contributions of the $B -$ and, particularly, the C-type of route-connecting pairs. For example for $M = 3$:

$$S = 0:\quad A^2 \to AB \overset{\nearrow\ AC\ \searrow}{\underset{\searrow\ B^2\ \nearrow}{}} BC \to C^2$$

$$S = 1:\quad A^2B \overset{\nearrow\ A^2C\ \searrow}{\underset{\searrow\ B^3\ \nearrow}{}} B^2C \to BC^2 \to C^2$$
$$B_2C^2 \nearrow$$

5. For a constant number of routes and intermediates, as well

Table 2.5. Complexity index k for different types and classes of kinetic graphs

No	code M-S/class	Complexity index K		
		N = 4	N = 5	N = 6
1	2	4	5	6
	2-0			
1	A	128	290	552-612
2	B	184	380-420	684-804
3	C	216-240	430-510	756-996
	3-0			
4	A^2	–	–	864
5	AB	–	600	1272
6	AC	–	750	1488-1680
7	B^2	384	880	1680-1872
8	BC	480	1030-1160	1896-2472
9	C^2	576	1180-1380	2112-3072
	3-1			
10	A^2B	–	600	1272
11	A^2C	304	690-750	1308-1680
12	B^3	384	880	1680-1872
13	B^2C	444-480	910-1160	1626-2472
14	B_2C^2	504-584	1000-1480	1752-2520
15	BC^2	576-620	1180-1510	2112-3072
16	C^3	768	1740	3312-3594
	4-0			
17	B^3	–	1600	3408
18	B^2C	–	2000	3988-4512
19	BC^2	–	2500	4980-5640
20	C^3	–	–	6360
	4-1			
21	B^3	–	–	3408
22	B^2C	–	–	3984
23	BC^2	–	–	4560
24	C^3	–	–	5136

Table 2.5. *Continued*

No	code M-S/class	Complexity index K		
		N = 4	N = 5	N = 6
1	2	4	5	6
	4-3			
25	BCB^3	800	1830–2000	3480–4512
26	$BCBC^2$	1312	3180	5136–7152
27	BC^3B	1376	3090–3370	5832–7680
28	C^5	–	–	10560
	4-4			
29	B^6	–	1600	3408
30	B^4C^2	1056	2400–2590	4560–5992
31	$B^2C^2B^2$	1000	2140–2420	3924–5640
32	B^2CBC^2	1328	2840–3290	5208–7560
33	B^2C^2BC	1396	3090–3360	5832–7680
34	BC^4B	–	4440	9228
35	$B_2^2C^2B_2C$	944–1120	1860–2420	3240–4872

as for a constant reaction type and class, K increases on going between the first subclasses I or L:

$$\ldots A_3 \to A_2 \to A \,;\, C \to C_2 \to C_3 \ldots$$

The transitions between the second subclasses K leave the complexity index unchanged.

6. For a constant number of routes and intermediates, and for a constant type, class, first and second subclass, K increases with a more uniform distribution of intermediates over the reaction routes (For example, for the mechanism with code $2-6-A-2-4$, $K = 552$ whereas for $2-6-A-3,3,C = 612$).

7. For a constant number of routes and intermediates, the weak linkage of two routes via 1 or 0 common intermediates (i.e. the linkage of the respective pair of cycles in the kinetic graph by means of a common vertex or a bridge, respectively) for classes containing

A^x, B^y, or $A^x B^y$ constituents results in the appearance of mechanisms of the same complexity (*isocomplicated mechanisms*). Such mechanisms may belong to different classes and subclasses (see paragraph 3) or differ solely in terms of the relative location of certain cycles (routes).

Another approach to the evaluation of reaction mechanism complexity that deserves attention is also based on algorithm (1), eq. (1) or its variants obtained by Yablonskii et al.[79] which can be transformed[32] into equation (4):

$$r_p = \frac{[X_i](r_p^y \cdot D_{po} + r_p^*)}{D_i} \tag{4}$$

where $r_p^y = \vec{C}_{po} - \overleftarrow{C}_{po}$ is the cyclic characteristic of route p:

$$r_p^* = \sum_{k=1}^{k \leqslant M-1} (\vec{C}_{pk} - \overleftarrow{C}_{pk}) D_{pk}, \tag{4'}$$

D_{po} characterizes the subgraph remaining from the kinetic graph after eliminating the cycle corresponding to route p; r^* is the conjugation factor (after Yablonskii) characterizing the degree of connectedness of a given cycle with respect to the remaining ones, and depends heavily on the kinetic graph class. Thus, for all the KGs of class B^y, $r^* = 0$, while the number of elementary step weights contributing to the term $\Sigma_{p=1}^M r_p^*$ in the case of class $4\text{-}3\text{-}C^5$ is 408.

Related to the foregoing, it would appear reasonable to evaluate the mechanism complexity by making use of the sum of the step weights included in D_{po} and r_p^* for all routes. This approach, however, necessitates a more detailed analysis.

2.5.3 Complexity of Mechanisms with Pendant Vertices

The complexity index can certainly be calculated for graphs with pendant vertices. In this case one takes into account the fact that the pendant vertices do not change the mechanism complexity because the basic graph topology (number of cycles, number of vertices in the graph's cyclic part, the cycle connectedness, etc.) is left unchanged. The presence of pendant vertices becomes of importance in the case of nonstationary processes. For stationary (or quasistationary) reactions having linear mechanisms, the pendant vertices influence the kinetic model, and in particular the material balance with respect to the catalyst (for noncatalytic reactions the change is with respect to the reagents).

When one uses eqn. (1) to determine the reaction date for routes (in those cases where the concentration of the active sites or complexes involved in the reaction can be measured[58]) and to calculate the K index, it is readily apparent that the addition of pendant vertices will not alter K. On the other hand, when the K index is calculated by means of eqn. (3) taking into account all the vertices and weights of all the elementary steps, the complexity index becomes unrealistically high. The contribution of the pendant vertices to the K index can be reduced if the latter is calculated from the general equation for the stationary rate for a route accounting for the pendant vertices[80]. The numerator of this equation is equal to the numerator of eqns. (1) and (2) while the denominator is specified by eqn. (5):

$$\sum_{i=1}^{N+N_p} D_i = \sum_{i=1}^{N} D_i F_i, \tag{5}$$

where N_p is the total number of pendant vertices, D_i is the determinant of vertex i neglecting the pendant vertices, and F_i is the so-called *complexity* of vertex i:

$$F_i = 1 + \sum_{l=1}^{l} \frac{w_{il}}{w_{-il}}. \tag{6}$$

Here w_{il} and w_{-il} are the weights of the forward and reverse oriented edges (arcs) proceeding from vertex i of the kinetic graph to the pendant vertex l.

Under steady-state conditions,

$$F_i = 1 + \sum K_{il} \prod C_j^{v_j} \tag{7}$$

For instance, $F_i = 1 + \Sigma b_{il} P_l$ for heterogeneous catalytic reactions on uniform surfaces, where b_{il} is the adsorption coefficient of reagent l at site i, and P_1 is the partial pressure of reagent l.

According to eqn. (5), each tree of vertex i connected to P_i pendant vertices will be repeated P_i times. Hence, one can introduce the quantity ΔK representing the increase in the complexity index K when including or neglecting the pendant vertices:

$$\Delta K = M(N-1) \sum_{i=1}^{N} T_i P_i, \tag{8}$$

where T_i is the number of spanning trees in ith vertex of the kinetic graph neglecting the pendant vertices.

Example: $M = 1, N = 3, N_p = 3, n_1(2),$ and $n_2(1),$
                (see the code of kinetic graphs with pendant vertices)
                $K = 1.2.9 + 6 = 24$
                $\Delta k = 1.2(3.2 + 3\ .1) = 18$
                $K = C + \Delta C = 42$

(Note: If all the spanning trees of all the graph vertices are taken into account one obtains $K = 102$).

2.5.4 Complexity of a System of Differential Kinetic Equations

Let us consider now the other type of kinetic model, namely the system of differential equations, and the ways of utilizing this model for evaluating the complexity of linear and non-linear mechanisms. The system of kinetic differential equations is given in matrix form by eqn. (9) which, for simplicity, is considered to be that for a closed system at constant volume:

$$\bar{C}_i^{\cdot} = B^T \times \bar{R}_k, \tag{9}$$

where $\bar{C}_i^{\cdot}$ is the vector column containing the derivative of concentration of all reagents with time, B^T is the transposed matrix from the stoichiometric coefficients for all the reagents in k elementary steps, and $\bar{R}_k$ is the vector column of the elementary step rates.

As we are interested in the structural information pertaining to the mechanism and the kinetic graph structure, it will be sufficient to deal only with the derivatives of the intermediate concentrations together with the respective simplified B matrix. We then obtain:

$$\bar{C}_x^{\cdot} = B_x^T \times \bar{R}_k, \tag{10}$$

where $\bar{C}_x^{\cdot}$ is the vector column of the derivatives with respect to time of the J concentrations for the linearly independent inter-mediates $x_i (N = J + 1)$, and B_x^T is the fragment of the $J \times K$ B^T matrix which includes the intermediates.

As follows from eqns. (9) and (10), the B_x^T matrix contains the entire structural information on the reaction mechanism (information on the graph structure). Each linear or non-linear mechanism can be characterized by means of this matrix.

Consider as an example a nonlinear mechanism containing two linearly independent intermediates, x_1 and x_2, four elementary steps, two reaction routes, and five elementary reactions with rates r_1, r_2, r_3, r_4, r_5 respectively.

$$\begin{vmatrix} \dfrac{dx_1}{d\tau} \\[2mm] \dfrac{dx_2}{d\tau} \end{vmatrix} = B_x^T \times \begin{vmatrix} r_1 \\ r_2 \\ r_3 \\ r_4 \\ r_5 \end{vmatrix}$$

$$\begin{matrix} B_x^T \\ (J \times K) \end{matrix} = \begin{Vmatrix} 2 & -2 & -1 & 0 & -1 \\ 0 & 0 & 2 & -1 & -1 \end{Vmatrix}$$

The B_x^T matrix will contain the following information:
(a) the number of elementary reactions K (and, respectively, the numbers of their weights, w_k);
(b) the number of linearly independent vertices and, therefore, the total number of graph vertices $(J + 1 = N)$;
(c) the number of reaction routes $M : M = K - J - \bar{K}$, where $\bar{K}$ is the number of reverse elementary reactions occurring in the B_x^T matrix;
(d) the stoichiometrix coefficient 2 stands for the reaction number and the number of the x_i taking part in a nonlinear elementary reaction; two entries of the same sign in a column also charac-terize an entirely nonlinear elementary step;
(e) in the case of linear mechanisms with reversible reaction steps, the number of nonzero entries in the row equals twice the degree of the respective vertex in the kinetic graph;

(f)   the magnitude of the nonzero entries in the rows and columns of B_x^T as well as the manner of their grouping, is specifically related to the graph structure.

Moreover, the B_x^T matrix also incorporates the pendant vertices in a natural way. In analysing the general form of the reaction mechanism, elementary steps resulting in the formation of "pendant" compounds should be regarded as nonequilibrium ones.

The information contained in the B_x^T matrix can serve as a basis for specifying a new complexity index which can describe both the linear and the nonlinear mechanisms of chemical reactions. The results obtained will be reported in a forthcoming publication[81].

2.6   Classification and Coding of Mechanisms applied to Kinetic Studies

As mentioned earlier, automatic systems for kinetic studies of chemical reactions have been developed intensively in recent times. The theoretical foundation for such systems involves a strategy based on the following scheme[8, 11, 82, 83]:

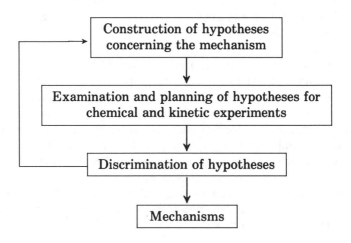

This strategy seems to be the only possible one for complicated reactions, even without automation of the individual stages. We now discuss the application of our system for the classification and coding of mechanisms in terms of each of the kinetic study stages shown in the scheme above.

2.6.1 Construction of hypotheses

Automation of the procedure for generating hypotheses by means of different formalized approaches[84, 8, 11, 14, 15, 85] requires the coding of mechanisms, as well as the organization of a mechanistic bank having a convenient system for information retrieval. The coding method we propose meets these requirements and can easily be automated. Thus, with only limited intervention of the researcher automated hypothesis generation can be carried out in parallel with automated coding.

We mention here that the necessity of naming mechanisms arose long ago in the theory and practice of complicated reaction kinetics studies. Such a notation (name) should take into account the mechanistic formal kinetic aspect only, reflecting the topological information it contains. The classification and code we propose for linear mechanisms solves this problem in an apt manner. Thus, for instance, the most typical mechanisms of heterogeneous catalytic reactions[8] are classified from a topological viewpoint (neglecting the pendant vertices and edge orientations) into the following types and classes (the notations used in the reference work[8] are given in brackets):

$$
\begin{array}{ll}
1 - 0 & (1.1;\ 4.1;\ 4.2;\ 6.1) \\
2 - C & (1.2;\ 3.1;\ 6.2) \\
2 - B & (2.1;\ 3.2;\ 4.3;\ 4.4) \\
3 - 1 - B^3 & (2.2) \\
3 - 1 - B_2 C^2 & (1.3) \\
3 - 1 - B^2 C & (5.1;\ 5.2)
\end{array}
$$

Any procedure for generating hypotheses is based on the simplicity principle. Because of this, we are justified in using the mechanistic hierarchy and in evaluating quantitatively the mechanistic complexity. It is instructive to examine "skeletal" schemes incorporating only intermediates without pendant vertices and having the minimum number of elementary steps which proceed on the basis of "one elementary step sequence – one product". The hypotheses thus generated can then be made gradually more complicated.

The topological approach proved to be of great importance in generating hypothetical multiroute mechanisms by combining one-route mechanisms[84].

2.6.2 Examination and discrimination of hypotheses

Each of the proposed hypotheses can be examined with the purpose of detecting those features that might be expected for the different classes of mechanisms. Kinetic equations are derived at this stage for the stationary reaction route rates and (in case of linear mechanisms) equations are obtained for the rate of substance formation or depletion or, in the most general case, systems of differential equations are written. Examination of the kinetic models allows us to devise a plan for model (hypothesis) discrimination using chemical, physicochemical, and kinetic methods. It is worthwhile discussing in more detail these methods for discriminating reaction mechanism hypotheses.

2.6.2.1 *Chemical and physicochemical discrimination*

Analysis of the chemical information contained in each hypothesis (such as the kind and stereochemistry of the intermediates, the mechanism of their interconversions, etc.) provides a basis for planning chemical and physicochemical experiments to reject a number of hypotheses. Such experiments include those using tracer molecules to examine feature distribution in the reagents, studies on the reagent conversions that are supposed to occur in the elementary steps of the process, inhibitor addition, radical trapping, the addition of process breaking in the different elementary steps and the detection of assumed intermediates by means of electrochemical, radiospectroscopic and other methods.

2.6.2.2 *Kinetic discrimination*

In performing these experiments it is important to decide which parameters need to be varied in the kinetic experiments, and to plan such experiments so that light is shed on the mechanism's topological structure or, more generally, to extract topological or structural information. It is also necessary to estimate qualitatively the connectedness of the mechanism graph, the location of the routes in the graph, their interconnections, as well as the presence of non-linear elementary steps.

Analysis of the kinetic equations of stationary processes reveals that only the mechanism of classes characterized by elementary steps common to two or more reaction routes ($C_{L,K}^Z$ – code substituent) differ in terms of their equation. Thus, for example, eqns. (4) for the rate for a route do not contain encompassing cycles when applied to classes A^x and B^y and, hence, $r_p^* = 0$. More specifically, for the 3-1-B^3-2,2,2 mechanism (graph 7), the catalytic reaction rate for route I is given

Figure 2.26

by the equation:

$$r_I = \frac{[M]_\Sigma (\vec{C}_{10} - \overleftarrow{C}_{10}) D_{10}}{\sum D_i} = \frac{[M]_\Sigma r_{10}^y D_{10}}{\sum D_i}, \tag{12}$$

where $[M]_\Sigma$ is the total catalyst concentration. Writing eqn. (12) in terms of the "free" catalyst concentration $[M_1]$, i.e. using the concentration of the substance, e.g. at vertex 1 we arrive at the equation

$$r_I = \frac{[M_1] r_{10}^y}{D_I^1}, \tag{13}$$

where D_I^1 is the determinant of the subgraph formed by cycle I at vertex 1. This equation contains only route I step weights. Hence, the rate ratio of routes I and II is:

$$\frac{r_{\mathrm{I}}}{r_{\mathrm{II}}} = \frac{r_{10}^y \cdot D_{\mathrm{II}}^1}{r_{20}^y \cdot D_{\mathrm{I}}^1}. \tag{14}$$

In the case of mechanisms for the classes C, C^2 and C^3 the situation is different. The rate along each route contains the elementary step weights of all the other routes, even when the material balance with respect to the catalyst is not taken into account. In the latter case, one finds for the $A^x B^y C^z$ mixed classes that only those routes that have elementary step(s) in common (a C-type for the supergraph edge) appear as mutually connected. For all mechanisms with a $C_{L,K}^z$ code substituent, there exists a linear dependence between the rate for one of the routes (r_i) and the sum of the rates (r_Σ) for all other routes having at least one elementary step in common with the ith route[59, 86]:

$$\alpha r_i + \beta r_\Sigma = \gamma \tag{15}$$

One may vary the concentration of any reagent that is not included in route i ($\alpha, \beta,$ and γ are constants). This dependence could be used for kinetic discrimination if the number of key reagents coincides with the number of routes (or, in other words, if the route basis is also their stoichiometric basis[9]).

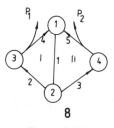

Figure 2.27

Analysis of the conjugation nodes in the consecutive elementary steps of formation of different products is the most general method kinetically discriminating among hypotheses (in the case of stoichiometrically non-unique reactions)[11]. For example, for the mechanism depicted by graph 8, the rate ratio for the elementary step of P_1 and P_2 product formation (r_I/r_{II}) is a simple function of the reagent concentrations included in the weights of the elementary steps 2 and 3:

$$\frac{r_{p1}}{r_{p2}} = \frac{w_2}{w_3} = \frac{K_2}{K_3} \cdot \frac{\Pi C_i^{\nu_i}}{\Pi C_j^{\nu_j}}, \tag{16}$$

and is independent of the weights of all the other elementary steps. Clearly, the possibility of checking experimentally on the existence of conjugation node (2) follows from the preceding analysis, provided steps 2 and 3 are not monomolecular ones and a method for measuring the r_{p1}/r_{p2} ratio exists. Some interesting examples of the use of conjugation node analysis can be found in studies on the mechanism of olefin metathesis[6, 87].

Chemical and kinetic discrimination among hypotheses is applied successfully to complicated conjugated homogeneous catalytic reactions involving the preparation of four esters from C_2H_2, CO, and ROH[11] (See Introduction). Proceding from the 1344 hypotheses generated under the assumption that each product is obtained from a single sequence of elementary steps, one arrives at only 90 mechanisms after chemical experiments and examination of the possible product interconversions. The remaining mechanisms have been analysed by juxtaposing the structure of the hypothetical conjugation nodes with the results of a single-factor kinetic experiment $r_i/r_j = \varphi(C_k)$. None of the 90 mechanisms was found to explain the whole collection of experimental data. For this reason, the procedure of hypothesis generation has been repeated assuming the possibility of product formation in several different sequences of elementary steps. By applying an analogous procedure for discriminating among the new set of hypotheses, four mechanisms have been found to agree with all the available data. One should bear in mind that dealing with such complicated mechanisms to find the explicit form of the dependence $r_i = f(c_i)$ is no solution to the problem. Thus, for example, by seeking a single-factor equation for the dependence of the acrylate synthesis rate on $[H^+]$, as well as using a relatively simple mechanism, an equation is obtained which includes a polynominal of degree 4 in [H$^+$] in the numerator and a polynominal of degree 5 in the denominator:

$$r_A = \frac{\sum\limits_{n=0}^{n=4} b_n [\text{H}^+]^n}{\sum\limits_{m=0}^{m=5} b_m [\text{H}^+]^m} \tag{17}$$

2.6.2.3 *Mathematical discrimination*

The hypotheses remaining after kinetic and chemical discrimination are subjected to statistical treatment taking into account their increasing complexity. With this purpose in mind, one needs complexity indices, i.e. criteria for evaluating the hypothesis complexity which enable one to order the hypotheses in some hierarchical sequence. Thereafter, the kinetic constants are estimated, the adequacy of the kinetic equations is checked, and a few more kinetic experiments are performed following a certain plan. For a large number of hypotheses it is natural to choose from the hierarchical sequence the simplest mechanism adequate for the experiment.

Bearing in mind the substaintial time consumption for the reverse problems in chemical kinetics, the disadvantages of the nonlinear statistical method of estimation,[88] and the known divergence of the solutions to such problems,[89] a definite preference may be accorded to the chemical and kinetic discrimination methods (as interpreted in the above, vide supra). They are related to the chemical and structural information for the studied reactions, and in practice they obviate the application of linear or nonlinear statistical methods of estimation for the model parameters. The application of nonstationary kinetics methods for obtaining information on the kinetics of the individual steps of the complex mechanism is very fruitful in solving the above-mentioned problems.

In Section 6 we have attempted to demonstrate the possible ways of using the various aspects of graph-theoretical classification and coding of the chemical reaction mechanisms in solving various problems in chemical kinetics, as well as in the development of computer simulation methods.

2.7 References

1. O.N. Temkin and R.M. Flid, Catalytic Conversions of Acetylene Compounds in Solutions of Metal Complexes (in Russian). Nauka, Moscow, 1968.
2. I.I. Moisseev, The π-Complexes in the Liquid Phase Oxidation of Olefins (in Russian). Nauka, Moscow, 1970.
3. G. Henrizi-Olivé and S. Olivé, Coordination and Catalysis. Verlag Chemie, Weinheim, N.Y., 1977.
4. P.M. Henry, Palldium Catalyzed Oxidation of Hydrocarbons, in "Catalysis by Metal Complexes", v. 2, ed. by B.R. James. Reidel, New York, 1980.
5. A.A. Klesov and I.V. Beresin, Enzyme Catalysis, Part I (in Russian), MGU, Moscow, 1980.
6. J.L. Bilhou, J.M. Basset, R. Mutin, and W.P. Graydon, J. Am. Chem. Soc. 99, 4083 (1977).
7. O.N. Temkin, S.M. Brailovskii, and L.G. Brouk, in Reports of All-Union Conference on Catalytic Reaction Mechanisms, vol. 1 (in Russian), Nauka, Moscow, 1978, p. 74.
8. Yu. S. Snagovskii and G.M. Ostrovskii, Modeling Kinetics of Heterogeneous Catalytic Processes (in Russian). Khimiya, Moscow, 1976.
9. S.L. Kiperman, Foundations of Chemical Kinetics of Heterogeneous Catalytic Processes (in Russian). Khimya, Moscow, 1979.
10. G.S. Yablonskii, V.I. Bykov, and V.I. Elokhin, Kinetics of Modeled Reactions in Heterogeneous Catalysis (in Russian). Nauka, Novosibirsk, 1984.
11. L.G. Brouk and O.N. Temkin, in Catalytic Reaction Mechanisms (III All-Union Conference), part II (in Russian). Novosibirsk, 1982, p. 10.
12. V.S. Shestakova, A.F. Kuperman, S.M. Brailovskii, and O.N. Temkin, Kinet. Katal. 22, 370 (1981).
13. O.N. Temkin and L.G. Brouk, Usp. Khimii, No 2, 206 (1983).
14. P.H. Sellers, Arch. Rational. Mech. Anal. 44, 23 (1971).
15. P.H. Sellers, in Chemical Applications of Topology and Graph Theory (R.B. King, Ed.). Elsevier, Amsterdam, 1983, pp. 420–429.

16. R. Barone, M. Chanon, and M.L. Green, J. Organomet. Chem. 185, 85 (1980).
17. I. Theodosiou, R. Barone, M. Chanon, J. Molec. Catalysis 32, 27 (1985).
18. Y. Yoneda, Bull. Chem. Soc. Japan 52, 8 (1979).
19. L.I. Shtokolo, L.G. Shteingauer, V.A. Likholobov, A.V. Fedotov, React. Kinet. Catal. Lett. 26, 227 (1984).
20. O.N. Temkin, L.G. Brouk, D. Bonchev, Teor. Exper. Khim. 24, 282 (1988).
21. M.I. Temkin, Dokl. Akad. Nauk SSSR 165, 615 (1965).
22. M.I. Temkin, in Mechanism and Kinetics of Complicated Reactions (S.Z. Roginskii, Ed.) (in Russian). Nauka, Moscow, 1970, p. 57.
23. A.I. Vol'pert, Matematicheskii Sbornik 88(130), 578 (1972).
24. A.T. Balaban, Ed., Chemical Application of Graph Theory. Academic, New York, 1976; Math. Chem. (MATCH) 1, 33 (1975).
25. D.H. Rouvray and A.T. Balaban, in Applications of Graph Theory, ed. by R.J. Wilson and L.W Beineke. Academic, London, 1979.
26. Chemical Applications of Topology and Graph Theory, ed. by R.B. King. Elsevier, New York, 1983.
27. Graph Theory and Topology in Chemistry, ed. by R.B. King and D.H. Rouvray, Elsevier, Amsterdam, 1987.
28. Discrete Appl. Math., ed. by J.F. Kennedy and L.V. Quintas, 19 (1988).
29. M.I. Stankevich, I.V. Stankevich, and N.S. Zefirov, Usp. Khimii 57, 337 (1988).
30. Computational Chemical Graph Theory, ed. by D.H. Rouvray, Nova Science, New York, 1990.
31. F.A. Feizhanov and V.A. Tulupov, Zh. Fiz. Khim. 53, 2261 (1979).
32. G.S. Yablonskii and V.I. Bikov, Dokl. Acad. Nauk SSSR 238, 645 (1978).
33. D. Bonchev, O.N. Temkin, D. Kamenski, React. Kinet. Catal. Lett. 19, 113 (1980).
34. D. Bonchev, D. Kamenski, O.N. Temkin, J. Comput, Chem. 3, 95 (1982).
35. O.N. Temkin, L.G. Brouk, and D. Bonchev, Teor. Exp. Khim. 24, 282 (1988).

36. O.N. Temkin and D. Bonchev, J. Chem. Educ. (submitted).
37. D. Bonchev, O.N. Temkin, D. Kamenski, React. Kinet. Catal. Lett. 19, 119 (1980).
38. D. Bonchev, D. Kamenski, and O.N. Temkin, J. Math. Chem. 1, 345 (1987).
39. H.L. Morgan, J. Chem. Docum. 5, 107 (1965).
40. M. Randić, J. Chem. Inf. Comput. Sci. 15, 105 (1975).
41. W.T. Wipke and T.M. Dyott, J. Am. Chem. Soc. 96, 4834 (1974).
42. C. Jochum and J. Gasteiger, J. Chem. Inf. Comput. Sci. 17, 113 (1977).
43. R.C. Read and R.S. Milner, Res. Rep, CORR 78-42 (1978), University of Warerloo, Ontario, R. Read, Res. Rep. CORR 80-7 (1980), University of Waterloo, Ontario.
44. D. Bonchev, O. Mekenyan, and A.T. Balaban, in Mathematics and Computational Concepts in Chemistry, ed. by N. Trinajstić, Ellys Horwood, Chichester (U.K.), 1986, p. 34.
45. Concepts and Applications of Molecular Similarity, M.A. Johnson and G.M. Maggiora, Eds., Wiley, New York, 1990.
46. D. Bonchev, A.T. Balaban, and O. Mekenyan, J. Chem. Inf. Comput. Sci. 20, 106 (1980).
47. D. Bonchev, A.T. Balaban, and M. Randić, Intern. J. Quantum Chem. 18, 369 (1980).
48. D. Bonchev, J. Mol. Struct. (THEOCHEM) 185, 155 (1989).
49. D. Bonchev, O. Mekenyan, and A.T. Balaban, J. Chem. Inf. Comput. Sci. 29, 91 (1989).
50. D. Bonchev and A.T. Balaban, J. Chem. Inf. Comput. Sci. 21, 223 (1981).
51. D. Bonchev, Pure Appl. Chem. 55, 221 (1983).
52. K. Gordeeva, D. Bonchev, O.N. Temkin, and D. Kamenski, (to be published).
53. I.S. Nagishkina and S.L. Kiperman, Kinet. Katal. 5, 7 (1964).
54. N.E. Bogdanchikova, G.K. Boreskov, P.A. Zhdan, G.Ya. Lastoushkina, and A.V. Hassin, in Reports of All-Union Conference on Catalytic Reaction Mechanisms, vol. 1 (in Russian). Nauka, Moscow, 1978, p. 267.
55. A.A. Khomenko, L.O. Apel'baum, F.S. Shub, Yu.S. Snagovskii, and M.I. Temkin, Kinet. Katal. 12, 423 (1971).
56. G.A. Kliger, A.M. Bashkirov, L.S. Glebov, O.A. Lessik, E.V. Marchevskaya, and P.A. Fridman, in Reports of All-Union Con-

ference on Catalytic Reaction Mechanisms, vol. 1 (in Russian). Nauka, Moscow, 1978, p. 251.

57. K.B. Yazimirskii, Application of Graph Methods to Chemistry. Naukova dumka, Kiev, 1971; Int. Chem. Eng. 15, 7 (1975).

58. S.M. Brailovskii, O.N. Temkin, and R.M. Flid, Kinet. Katal. 12, 1152 (1971).

59. S.M. Brailovskii, O.N. Temkin, and A.C. Kostyushin, Kinet. Katal. 31, 1371 (1990).

60. S.V. Pestrikov, I.I. Moiseev, and L.I. Sverzh, Kinet. Katal. 10, 74 (1969).

61. A.I. Gel'bshtein, M.G. Slin'ko, G.G. Shcheglova, G.S. Yablonskii, V.I. Timoshenko, and B.A. Kamenko, Kinet. Katal. 13, 709 (1972).

62 V.M. Emanuel and D.G. Knorre. Chemical Kinetics Course (in Russian). Visshaya shkola, Moscow, 1974, p. 319.

63. A.T. Balaban, I. Motoc, D. Bonchev, and O. Mekenyan, in Steric Effects in Drug Design, ed. by M. Charton and I. Motoc. Topics in Current Chemistry 114, Springer, Berlin, 1983, p. 21.

64. N. Trinajstić, Chemical Graph Theory, Vol. 2, CRC Press, Boca Raton, Florida, 1983, p. 105.

65. D. Bonchev, Information Theoretic Indices for Characterization of Chemical Structures. Research Studies Press, Chichester, 1983.

66. D.H. Rouvray, Sci. Amer. 254, 40 (1986).

67. D.H. Rouvray, J. Comput, Chem. 8, 470 (1987).

68. H. Wiener, J. Am. Chem. Soc. 69, 17 (1947).

69. D.H. Rouvray, in Mathematics and Computational Concepts in Chemistry, N. Trinajstić, Ed. Ellis Horwood, Chichester, 1987, p. 295.

70. A.T. Balaban, Chem. Phys. Lett. 89, 399 (1982).

71. S.H. Bertz, J. Am. Chem. Soc. 103, 3599 (1981).

72. D. Bonchev, O. Mekenyan, and N. Trinajstić, J. Comput. Chem. 2, 127 (1981).

73. S.H. Bertz, in Chemical Applications of Topology and Graph Theory, R.B. King, Ed. Elsevier, Amsterdam, 1983; p. 192.

74. S.H. Bertz and W.C. Herndon, in Artificial Intelligence Applications in Chemistry, T.H. Pierce, B.A. Horne, Eds. ACS Symposium Series 306, ACS, Washington, DC 1986, p. 169.

75. D. Bonchev and O.E. Polansky, in ref. 23, p. 126.
76. D. Bonchev, in ref. 30, Chapter 2, p. 34.
77. M.V. Vol'kenshtein and B.N. Gol'dshtein, Dokl. Akad. Nauk SSSR 170, 969 (1966).
78. M.V. Vol'kenshtein, Physics of Enzymes (in Russian), Nauka, Moscow, 1967, p. 155.
79. V.A. Evstigneev, G.S. Yablonskii, and V.I. Bykov, Dokl. Akad. Nauk SSSR, 245, 871 (1979).
80. L.G. Brouk and O.N. Temkin, in Reports of IV All-Union Conference "Mathematical Methods in Chemistry" (in Russian). Erevan, 1982, p. 230.
81. O.N. Temkin and D. Bonchev (in preparation).
82. V.N. Pissarenko and A.G. Pogorelov, Planning of Kinetic Studies (in Russian). Nauka, Moscow, 1969.
83. D. Kamenski, O.N. Temkin, and D. Bonchev, Appl. Catal. (submitted).
84. O.N. Temkin, S.M. Brailovskii, L.G. Brouk, in: Chemical Kinetics in Catalysis, USSR, Chernogolovka, 1985, p. 59 (in Russian).
85. V.I. Dimitrov, Simple Kinetics (in Russian). Nauka, Novosibirsk, 1982.
86. S.M. Brailovskii and O.N. Temkin, in Catalytic Reaction Mechanisms (IV All-Union Conference), Part II (in Russian). Novosibirsk, 1982, p. 97.
87. T.J. Katz, J.Mc. Ginnis, J. Am. Chem. Soc. 99, 1903 (1977).
88. V.V. Nalimov, Zavodskaya Laboratoriya 44, 325 (1978).
89. S.I. Spivak, in Mathematical Problems of Chemical Thermodynamics (in Russian). Nauka, Novosibirsk, 1980, p. 63.

Chapter 3

GRAPH THEORY AND THE MECHANISTIC DESCRIPTION OF CHEMICAL PROCESSES: SOME REMARKS ON AROMATICITY

P.J. Plath, E.C. Hass[1] and M. Kramer[2]

Institut für Ángewandte und Physikalische Chemie

Forschungsgruppe–Reaktionsmodelle, Universität Bremen,

Bibliothekstrasse, NW 2, D-2800 Bremen 33, Germany

[1]E.C. Hass

Spandau, Jagowstrasse 10, D-1000 Berlin 20, Germany

[2]M. Kramer

F.B. Chemie, Pädagogische Hochschule Halle, Kröllwitzer Strasse,

D(0)-4002 Halle, Germany

3.1 Introduction

The formulation of chemical processes via the description of molecular mechanisms is the basic procedure for gaining knowledge in, for instance, laboratory prognostic chemistry. Although there exists an extensive set of rules, which allow one to formulate molecular mechanisms for chemical reactions on the basis of chemical structure formulas, we still know comparatively little about the logic which governs these rules (1, 2).

Moreover, the theory of structure itself is not unequivocally defined in chemistry. The connection between the idea of chemical structure and the theory of dynamic systems is still vague (3, 4, 5).

Chemical formulas can be translated into mathematical language by using graph-theoretical structure theory, which enables us to deal with the problems arising from chemical structure patterns.

It is our aim to show, by using the typically chemical notion of aromaticity as an example, appropriate applications of the mathematical formalisms of graph theory in the solution of some of the problems mentioned above.

The notion of aromaticity is interesting for several reasons: it was with this idea that one extremely difficult problem of chemical structure theory was first solved: the description of the benzene molecule by two separate pictures, both of which are Kekulé formulas (7).

We know today and Kekulé knew then that the description of the benzene molecule by each picture on its own is incomplete, in that it describes only one of two possibilities as reality and not the interaction of the totality, which leads to different structures.

We have learned how to deal with these problems inherent in structure theory. A new symbol has been created to evade these contradictions, though the problem is not resolved – it is merely set aside. With the development of quantum mechanics the problem resurfaced, clothed in wave-particle dualism, though we were not conscious of the similarity of the problems.

In the formalisms of quantum mechanics the basic problem has been discussed. Today these problems are understood to the extent that we have at our disposal a formulation of the logic of quantum mechanics (8, 9).

One can ask whether these circumstances can be discussed only in the language of quantum mechanics together with its associated semantic implications. There is no doubt that even in other areas of science events can be observed in which their link can no longer be satisfactorily described in terms of Boolean logic.

In our opinion the idea of aromaticity in chemistry belongs to the class of ideas comprised of contradictory events.

Because of its generalization in respect to heterocyclic compounds and pericyclic reactions, however, the problems linked with the idea of aromaticity assume a general character, which goes much deeper than that for the historical example of benzene.

3.2 Graph-Theoretical Structure Theory

3.2.1 The Concept of Molecular Reaction Graphs (2, 6)

As historically substantiated and continually reconfirmed, the language of chemical formulas expresses the chemical connections of the atoms; its basic structure is therefore topological (1, 10, 11, 12). In classical chemical structure theory molecular connections are always expressed by the structural formula, whose elements are atoms and valencies. These very abstract ideas describe chemical occurrences. It has frequently been shown that these structural formulas can be expressed either as graphs or in topological form: they could be mapped on to simplicial complexes of dimension $C = 1$ (2, 13–16). Most of these attempts remained at the transition stage from chemical structural formulas into graph theory. They did not describe the reaction processes. In order to go one step further it is necessary to reflect upon the classical steps taken by chemists. Using structure theory, the chemist associates the qualitative changes substances undergo in chemical reactions with the changes in the structure of their molecular representatives.

1. By this method the atoms of a substance were combined to form a molecule using only valency links to construct the representative of the substance. This method can be used only when the connections between the atoms are established by chemical reactions (or spectroscopic data).
2. In chemistry the molecule is the "representative of a substance"; this idea is thus subject to the external condition that the number of atoms in the molecule has to be a minimum.

With the help of such extra conditions, which are alien to graph theory, models of chemical formula languages can be established, i.e. Graph-theoretical structure theory can be developed.

Therefore we can suppose that every molecular situation can be represented by a graph $G = (V, E)$ with a finite or infinite but maximal denumerable set of vertices $V_i (i \in I)$.*

The set E of edges of the graph G forms the structure of the molecule, which is complete for chemists. The idea of structure as used here is understood in a general way as a set of subsets of the set of elements. The idea of a molecule, as elsewhere in chemistry, can be established only by chemical reactions. The idea holds only when valencies are altered through formation or elimination during one possible reaction.

In the framework of graph-theoretical structure theory, the chemical process itself is given a structure determined by the changing of molecular arrangements.

In other words, the sets E_D of the changing edges between the graphs mapped on to the reactants give the structure of chemical processes (2, 18–22). Since the structure E_D is operating on the same set of vertices $V_D = V$ as the graph $G = (V, E)$ of

---

*Remark:

Given a real subgraph $G' \subset G$ it should be finite: G could also be a hypergraph. (17) In this case the circumference m of the edge must be finite. The circumference m of the edge m_k is the number of vertices incident to this edge.

the molecular situation representing the reactants or products, the structure E_D can be viewed as the set of edges of a graph $D = (V_D, E_D) = (V, E_D)$.

The graph D is called the dynamic graph of the reaction. The character of a chemical reaction is given by the structure E_D of D, as shown in the example of pericyclic reactions, which are distinguished by the fact that all the D graphs are cycles (see Figure 3.1).

Normally, in a given reaction, not all edges will be changed at the same time. The edges remaining constant during a reaction form the invariant part E_S of the whole structure E_M of the reaction. They first become evident in other structures of reactions. The set of all edges of D and all invariant edges E_S form the set of edges in graph M. This graph is the structure-theoretical picture of the reaction; it is termed the molecular reaction graph M. It describes the same facts otherwise described by the transition state or transition complex (23).

The difference between the constant and variable parts of the structures in the reaction is similar to the analysis of the set of edges M in the sets of static and dynamic edges. These are related to the static graph $S = (V, E_S)$ with the same set of vertices $V_S = V$, and with the set of dynamic edges to the graph $D = (V, E_D)$-again with the same set of vertices $V_D = V$.

S is therefore a subgraph of the graphs of the products A_p and reactants A_E and may be called a skeleton of the reaction (see fig. 3.2).

The static graph S is defined only in regard to the observed reaction. The orthocomplement of S in respect to M is the graph D, while M represents the milieu of D and S. Although the graph S of M is never an empty graph ("O" graph), its set of edges E_S may be an empty set $E_S = \phi$.

As shown in the example of pericyclic six centered reactions, the dynamic graph D consists of C_6 graphs in all cases and the graph S provides for the classification of reactions (see Figure 3.1) (18, 24–26).

It should be mentioned that the behavior of molecules, known as mesomerism, can be interpreted formally as a reaction.

In benzene-like systems, aromaticity emerges if $S = D = C_n$

Classification of Pericyclic Reactions

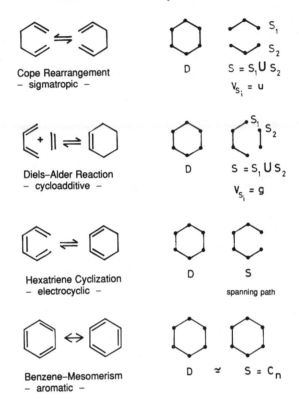

Figure 3.1. Classification of even membered pericyclic reactions.

holds. At this level one cannot distinguish between different forms of aromaticity.

3.3 The Topology of Pericyclic Reactions

The binary nature of the connections within graph-theoretical structure theory raises the issue as to the underlying algebra. We may ask what algebraic structure would be able to analyze the path

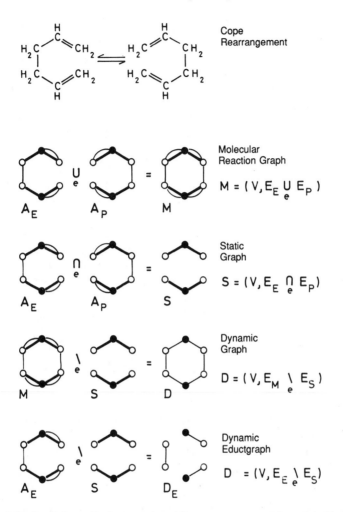

Figure 3.2. Graph-theoretical expression of Cope rearrangement formulated in terms of chemical structure theory. $A_E = (V, E_E)$ graph of reactants; $A_p = (V, E_p)$ graph of products. The binary graph-theoretical linkings operate only on the edge sets.

of the chemical reaction and provide the ideas we need to describe the reaction.

To investigate this issue, we start from the concept of the molecular reaction graph, and particularly the binary relations defined within this framework, i.e. the union $\cup_e$ and the intersection $\cap_e$ and the set-theoretical difference $_e\backslash$ of the edge sets of two graphs.

For example, if we consider the cyclization of butadiene to yield cyclobutene, the graphs A_E, A_p, M, S, D, D_E and D_p together with the graph containing no edges form a closed totality with respect to the operations mentioned above. Note that these operations are defined only on the edge sets of the graphs and that all unions and intersections of edges are subsets of the set containing *all* edge sets and that both operations are reflexive. As a consequence, the set of all edge sets of these graphs forms a topology which in particular is isomorphic to the discrete topology of the set consisting of three elements on the power set $P(3)$ of three numbers, respectively (see Table 3.1). Without loss of generality, the edges can also be indexed in such a way that both graphs S and D contain only edges with the

Table 3.1. The set of edge sets isomorphic to the power set P(3) belonging to the pericyclic isomerization of butadiene to cyclobutene

$$V = \{1, 2, 3, 4\}$$
$$A_E = (V, \{\{1, 2\}_1, \{1, 2\}_2, \{2, 3\}_1, \{3, 4\}_1, \{3, 4\}_2\})$$
$$A_P = (V, \{\{1, 2\}_1, \{2, 3\}_1, \{2, 3\}_2, \{3, 4\}_1, \{1, 4\}_1\})$$
$$M = (V, E(A_E) \cup_e E(A_P))$$
$$= (V, \{\{1, 2\}_1, \{1, 2\}_2, \{2, 3\}_1, \{2, 3\}_2, \{3, 4\}_1, \{3, 4\}_2, (1, 4\}_1\})$$
$$S = (V, E(A_E) \cap_e E(A_P))$$
$$= (V, \{\{1, 2\}_1, \{2, 3\}_1, \{3, 4\}_1\})$$
$$D = (V, E(M) \underset{e}{\backslash} E(S))$$
$$= (V, \{\{1, 2\}_2, \{2, 3\}_2, \{3, 4\}_2, \{1, 4\}_1\})$$
$$D_E = (V, E(A_E) \underset{e}{\backslash} E(S)) = (V, E(A_E) \cap_e E(D))$$
$$= (V, \{\{1, 2\}_2, \{3, 4\}_2\})$$
$$D_P = (V, E(A_P) \underset{e}{\backslash} E(S)) = (V, E(A_p) \cap_e E(D))$$
$$= (V, \{\{2, 3\}_2, \{1, 4\}_1\})$$
$$\Phi_4 = (V, \Phi)$$

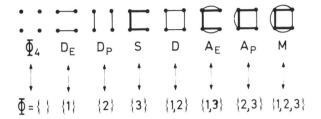

Φ_4 D_E D_P S D A_E A_P M

$\Phi = \{ \}$ $\{1\}$ $\{2\}$ $\{3\}$ $\{1,2\}$ $\{1,3\}$ $\{2,3\}$ $\{1,2,3\}$

Figure 3.3. The one-to-one mapping of graphs of the reaction and elements of the discrete topology of the power set $P(M)$ with $M = (1, 2, 3)$. The edges of S and D are distinguished by light and bold lines, representing different indicators.

same indices, respectively. The set of edge sets of graphs S, D_E and D_p forms the set of elements on which the discrete topology is defined as a system of subsets (see Figure 3.3).

Analogously to the example mentioned above, the other classes of pericyclic reactions with even numbers of chemical centers can be related to isomorphic topologies of edges. The structure does not change if we increase the number of atoms taking part during these reactions. The topologies of pericyclic reactions are equal up to isomorphism.

3.4 Lattices of Pericyclic Reactions

Starting from the fact that the edge sets of graphs describing a pericyclic reaction of an even number of vertices of their graphs form a topology isomorphic to the power set $P(3)$, the lattice that can be mapped on to this set has to be Boolean with respect to the inclusion relation (see Figure 3.4). Since it is Boolean, it is atomic and uniquely complemented. Each type of pericyclic reaction can now be described by a special Boolean reaction lattice, which in turn is isomorphic to the lattice $L(M) = P(3)$ if the dynamic graph D consists of an even number of vertices (see Figure 3.5).

The diagram of those lattices corresponds to a cube. The particular pericyclic reactions are distinguished by the edge sets of the set-theoretical atoms D_E, D_p and S. But the lattices of all the

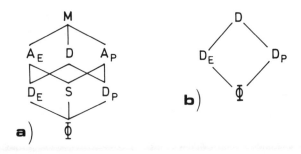

Figure 3.4. (a) Generalized three-dimensional reaction lattice and (b) its two-dimensional dynamic sublattice.

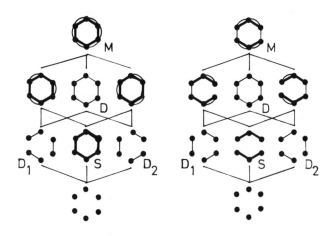

Figure 3.5. Reaction lattices of benzene-like aromaticity and of the sigmatropic Cope rearrangement.

pericyclic reactions contain the same "dynamic sublattice" L_D (see Figure 3.4) if it is Boolean and its diagram is isomorphic to that of a square (two dimensional cube).

In the sense of lattice theory, pericyclic reactions with an even number of chemical centers will differ essentially from those with an odd number of centers. In the latter case we restrict ourselves to the description of monocyclic molecular systems. Whether we look at neutral (and thus radical), ionic, e.g. anionic or cationic molecular structures has no bearing on the fact that there exist more than two Kekulé-formulas in any case (see Figure 3.6).

The appearance of nonbonded electrons or pairs of electrons lead us to introduce, respectively, directed or undirected loops into the corresponding graphs. Following this procedure in constructing reaction graphs, it can be shown that the graphs D_i do not need to be disjoint with respect to their set of edges (see Figure 3.7). In such a case they form a subbasis of a topology only, together with the set of edges of the graph S, but not a basis as in the case of the disjoint set of edges of the graphs D_i. The partially ordered set (= poset) of the set of sets of edges generated by the inclusion relation is no longer a lattice[+] but a poset[+]. A part from the individual concrete interpretation of the elements of these posets, the structure of the

---

[+]Remark on (28): A lattice $L = (L, \wedge, \vee)$ is an ordered set L, which has a *lower bound* and an *upper bound* for any two elements. In any ordered set in which there exist a lower and an upper bound we have the identities:

Idempotent Law $x \wedge x = x$ $x \vee x = x$
Commutative Law $x \wedge y = y \wedge x$ $x \vee y = y \vee x$
Associative Law $(x \wedge y) \wedge z =$ $(x \vee y) \vee z =$
                             $x \wedge (y \wedge z)$ $x \vee (y \vee z)$
Absorption Law $x \wedge (x \wedge y) = x$ $x \vee (x \vee y) = x$

A poset is an ordered set with an idempotent, commutative and associative binary operation. If P is an ordered set in which there is defined a *meet* denoted by $\wedge$ for any two elements a and b: $a \wedge b$; $P = (P, \wedge)$ is called a poset; $P = (P, \vee)$ is a poset in any case.

[++]Remark on (29): A non-empty finite partial order H can be represented by a diagram. If a, b, are elements of H, and b is the upper neighbor of a, the point (vertex) P_b corresponding to b is plotted above the point P_a corresponding to a (lateral shifting is allowed). P_b and P_a are connected by a line segment.

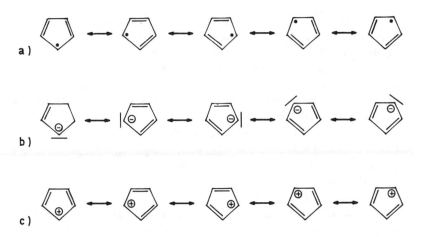

a)

b)

c)

Figure 3.6. Mesomeric Kekulé structures of the (a) cyclopentadienyl radical, (b) cyclopentadienyl anion, and (c) cyclopentadienyl cation.

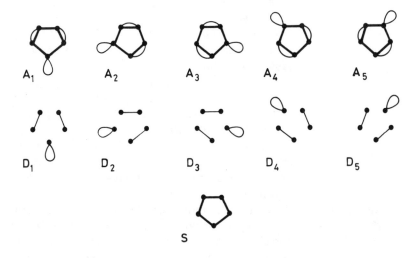

A_1 A_2 A_3 A_4 A_5

D_1 D_2 D_3 D_4 D_5

S

Figure 3.7. Graphs A_i of mesomeric Kekulé structures of the cyclopentadienyl cation and its corresponding dynamic subgraphs D_i and its static graph S.

underlying diagram[++] is similar to that of a Boolean lattice with a corresponding dimension.

Such posets can also be obtained by describing condensed aromatic compounds in terms of their Kekulé structures (see Figure 3.8).

This structure of the lattices reflects the remarkable fact that the properties of the system as a whole in the case of conjugated odd numbered monocyclic rings as well as in the case of condensed aromatics are not deducible by addition of the properties of the extreme Kekulé structures, since each of these limiting structures partially contains others.

In conventional terms, mesomerism in pyrrole and naphthalene are not reactions at all, and one might be forced to a high dimension of the reaction lattice by the special situation of mesomerism, although the counterexample of benzene aromaticity exists.

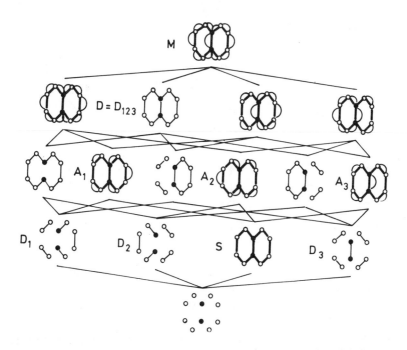

Figure 3.8. Semilattice of the mesomeric Kekulé structure of naphthalene.

For this purpose we mention a pericyclic reaction which may be assumed to pass through real ionic intermediates because of special substituted groups (see Figure 3.9)(30).

The structure-theoretical analysis of this reaction again gives three dynamical graphs D_i ($i = 1, 2, 3$) (Figure 3.3). Together with the graph $S \simeq P_2 \cup P_2$, these graphs form the four atoms of the four-dimensional Boolean reaction lattice, which again contains a three-dimensional Boolean dynamic sublattice.

However, the pericyclic reactions with an odd number of centers possess a special feature of importance with respect to aromaticity.

The union of the set of edges of all graphs D_i does not have to be a simple cyclic graph only but can contain loops too. In this sense we need to expand the former definition of aromaticity.

A system is called aromatic if S and D are cycles isomorphic to

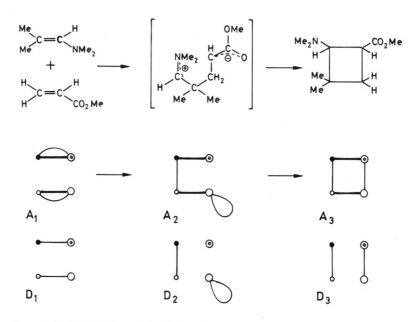

Figure 3.9. Cycloaddition with a double ionic intermediate: the corresponding graphs A_i of the reactants (i) and their contained dynamic subgraphs D_i.

each other up to the loops (5). However, there is no need to take all graphs D_i to create a graph containing a cycle. It is sufficient to insist that one needs only a minimal set of graphs D_i to form a cyclic graph (Figure 3.10).

But there is more than one of these minimal systems of generating graphs D_i. In the case of five-membered rings, one needs at least three graphs D_i, having no disjoint set of edges.

The incorporation of heteroatoms into the ring system does not change the former statements essentially; this corresponds to the coloring of knots of graphs.

As the example of pyridine shows, it can happen that the static graph S contains loops too (Figure 3.11).

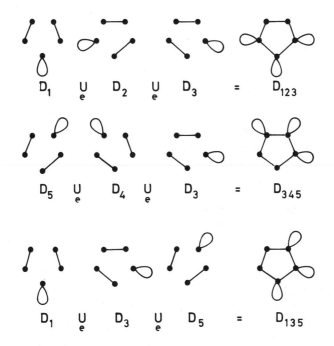

Figure 3.10. Minimal sets of graphs D_i needed to contruct a dynamic graph D_i' containing a cycle as a subgraph. The example represents the cyclopentadienyl anion.

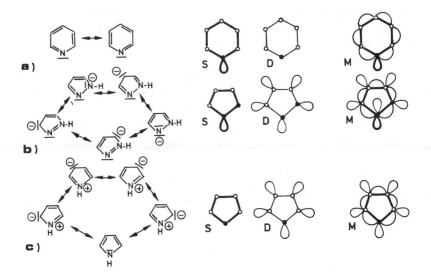

Figure 3.11. Mesomeric Kekulé structures of various systems: (a) pyridine, (b) pyrazole, (c) pyrrole and their graphs *S*, *D* and *M*.

In case of neutral ring systems with an odd number of centers there exists only one Kekulé structure without charge separation, in contrast to neutral and even membered rings.

3.5 Generalized Reaction Lattice

Whether we have a benzene-like system, catalyzed or non-catalyzed pericyclic reactions with an even number of centers, or the Demjanow-reaction, structure-theoretical analysis always results in a corresponding m-dimensional reaction lattice containing an $(m - 1)$-dimensional, dynamic sub-lattice, which is also Boolean (see Figures 3.12 and 3.13).

In the cases of pericyclic reactions with an odd number of centers and condensed aromatic compounds, we obtain a poset (L, v) with respect to the union of edge sets. Its underlying abstract diagram is isomorphic to the diagram of a Boolean lattice, but its

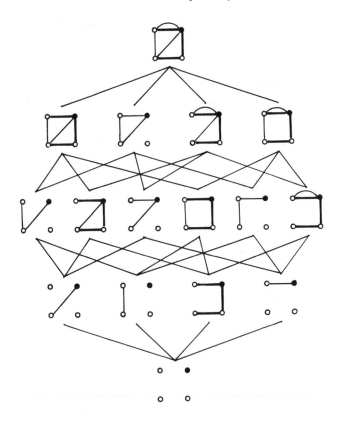

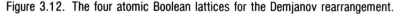

Figure 3.12. The four atomic Boolean lattices for the Demjanov rearrangement.

elements cannot be interpreted in a concrete manner without loosing the characteristic properties of a lattice and forming a poset.

Starting from the non-interpreted diagram, which corresponds to a Boolean-lattice in any case, one can label each level of this diagram by a dimension Δ.

In doing so, one makes use of the analogy between the number of elements of each level of this diagram and Euler's theorem for polyhedra (27).

The level of the lattice-theoretical atoms is labelled by the dimension $\Delta = 0$ (abbreviated to Δ^0)(see Figure 3.14).

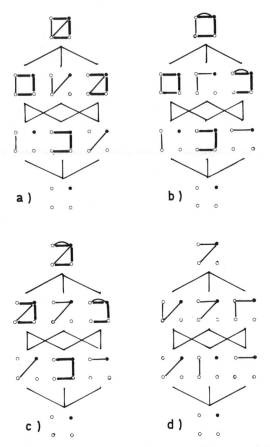

Figure 3.13. The four three-dimensional Boolean sublattices of the Demjanov rearrangement which can be interpreted "mechanistically": (a) cyclobutyl-/cyclopropylcarbinyl rearrangement (b) cyclobutyl-allylcarbinyl rearrangement (c) cyclopropylcarbinyl allylcarbinyl rearrangement (d) describes the "non-mechanistic" or "non-classical" character of the transition by a dynamic sublattice.

The connection between any two such atoms by a binary operation on the union of edges leads to an element belonging to the level with the dimension $\Delta = 1$. The idea of dimension introduced here seems to be related to the idea of the degree of a term in the sense of predicate logic.

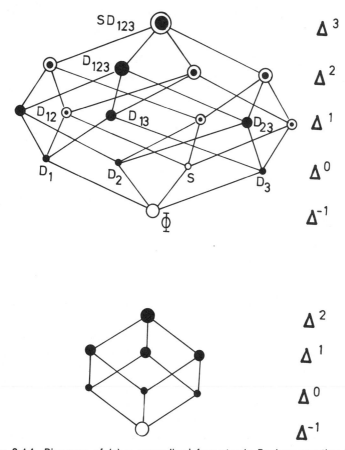

Figure 3.14. Diagrams of (a) a generalized four atomic Boolean reaction lattice and (b) its dynamic sublattice. Labelling of the different levels of the diagram by a dimension Δ^m.

In the framework of classical structure theory, a reaction can be described as a *mechanistic reaction* if there exists only one element $D_{max} = D = \cup_i D_i$ of the dynamic sublattice that can be labeled by the dimensiion $\Delta = 1$. However, if a reaction lattice has to be formed and D_{max} has to be labeled by a dimension $\Delta > 1$, one cannot show a mechanism for this reaction or the statement of a mechanism is no longer unique.

However, in such a case one can find sublattices of the reaction lattice which can be interpreted mechanistically (see Figure 3.13).

Within the framework of this argumentation, based on the idea of dimension, it is meaningless to ask whether the chemical graphs associated with the atoms of the abstract lattice are disjoint or not.

On the other hand, we are convinced that the difficulties arising from the transfer of the idea of aromaticity to conjugated, odd-numbered, monocyclic rings and condensed aromatic compounds are caused by the fact that atoms of the abstract diagrams chemically interpreted by dynamic graphs have no disjoint edge sets and so create posets rather than lattices.

3.6 The principle of valency conservation (λ-model) – An algebraic approach to the description of pericyclic reactions (2, 22, 31)

Starting from the discrete logical structure of the reaction lattice, let us describe a chemical reaction algebraically by employing continuous parameters. In the simplest case we shall investigate the dynamic sublattice of the three-dimensional reaction lattice (single parameter λ-model). Thereafter, these basic ideas will be generalized in three ways:

1) extension of the formalism to dynamic sublattices of arbitrary dimensions (multi-dimensional λ-model);
2) inclusion of the static graph into the algebraic treatment (i.e. reaction invariant influences on the dynamic system will be considered)
3) description of heteroatom influences by "coloring" D and S.

3.6.1 The single parameter λ-model

A pericyclic reaction involving only reactants and products as reaction partners and described by a two-dimensional Boolean dynamic sublattice may be interpreted in a statistical way by employing arguments based on orthomodular lattices (58, 59). For this purpose the dynamic subgraphs D_1 and $D_2 = D_1^\perp$, which are complementary graphs with respect to D, are conceived of as negating each other (6, 32). Introducing a continuous para-

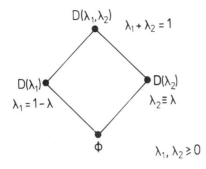

Figure 3.15. Normed, dynamic Boolean sublattice.

meter space with D_1 and D_2 as boundaries, any "transition situation" between the complements D_1 and D_2 may be described by oppositely weighting their edges with $(1 - \lambda)$ and λ ($\lambda \in [0, 1]$), respectively.

In this way the dynamic Boolean sublattice is normalized (Figure 3.15), and this in turn allows a probabilistic treatment (ref. 33) of the transition between D_1 and D_2 and thus of the dynamic aspect of the reaction considered (single-parameter λ-model). In other words, based on a statistical interpretation, any value of the "reaction parameter" λ corresponds to a structural mixture $D(\lambda)$ consisting of the structures D_1 and D_2 with fractions $(1 - \lambda)$ and λ, respectively. Thus, in chemical terms, any λ-value characterizes a situation in a pericyclic reaction where the "reacting" bonds of the reactants are reduced by the factor $(1 - \lambda)$ and, simultaneously, those of the products are newly formed to the extent λ. During the whole reaction the sum of all fractional bonds is invariant, i.e. the principle of valency conservation holds. The λ-model can thus be viewed as an extension of this principle – well known in chemistry for the reactants and products – to the overall chemical reaction. For the purposes of algebraic treatment, the λ-weighted graphs D_λ of the normalized dynamic sublattice are transformed into their adjacency matrices $\underline{A}(D_\lambda)$ (more precisely matrices of edge values) (34).

In the case of a pericyclic four-center reaction via a Hückel-type transition complex (11) one obtains for example:

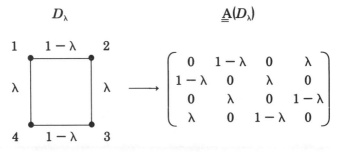

In general, the set of λ-weighted dynamic graphs of a pericyclic reaction can be expressed by the matrix function:

$$\underline{\underline{A}}(D_\lambda) = \underline{\underline{A}}_\lambda(D) = (1 - \lambda)\underline{\underline{A}}(D_1) + \lambda\underline{\underline{A}}(D_2) \tag{1}$$

with $\underline{\underline{A}}(D_1)$ and $\underline{\underline{A}}(D_2)$ having equal adjacencies D_1 and D_2, respectively (2). $\underline{\underline{A}}_\lambda(D)$, i.e. the λ-dependent matrix of edge weights of $D(\lambda)$, can be interpreted as a uniform description of the continuous aspect of the reaction to be considered.

In analogy to simple Hückel molecular orbital theory (HMO-theory (35–37) based essentially on the connectivity of the respective molecule, and very successfully applied to an understanding of conjugated polymers (10, 38, 39), the eigenvalue spectra of the dynamic graphs D_λ and their corresponding matrix functions $\underline{\underline{A}}_\lambda(D)$ may be interpreted "energetically" and, hence, correlated with the stability properties of the transition. In the following, we focus on the method of getting and interpreting eigenvalue correlation diagrams, whereas in a later chapter we shall attempt to give a theoretical explanation for the formalism by employing a stability analysis of graph spectra.

The λ-dependent eigenvalue of the matrix function $\underline{\underline{A}}_\lambda(D)$ is given by (40):

$$\underline{\underline{A}}_\lambda(D)\underline{\underline{C}}_\lambda(D) = \underline{\underline{C}}_\lambda(D)\underline{\underline{\Gamma}}_\lambda(D) \tag{2}$$

with $\underline{\underline{\Gamma}}_\lambda(D) = \text{Diag}\,(\gamma_k)$ equal to the matrix of eigenvalues in diagonal form and $\underline{\underline{C}}_\lambda(D)$ equal to the corresponding eigenvector matrix. Drawing all eigenvalue correlation lines $\gamma_k(\lambda)$ $(k = 1, 2, \ldots, n;$ $n = $ number of vertices of D_λ) as a function of the variation parameter λ within the boundaries 0 and 1 yields a "graph-theoretical eigenvalue correlation diagram" (22).

Figure 3.16 shows the correlation diagram for the pericyclic four-center reaction via a Hückel-type transition complex as mentioned above. It contains two crossing correlation lines (γ_{2+} und γ_{2-}), which are indicative of a thermally forbidden transition (2). Examples for such a reaction via an antiaromatic four-membered Hückel ring are the dimerization of ethylene and the disrotatory ring opening of cyclobutene (41), both of which occur only photochemically but not thermally.

On the other hand, the conrotatory ring opening reaction of cyclobutene is thermally allowed, and occurs via a Möbius-like (43) cyclic transition complex according to Dewar (11) and Zimmermann

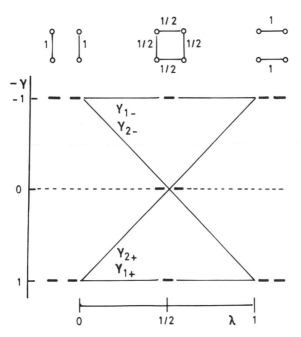

Figure 3.16. Eigenvalue correlation diagram of pericyclic four-center reactions via Hückel-type transition state.

(42). In the framework of graph-theoretical structure theory as well as the single parameter λ-model and using Möbius graphs (44) such a reaction can be described by the following D_λ or $\underline{\underline{A}}(D_\lambda)$ (22) thus:

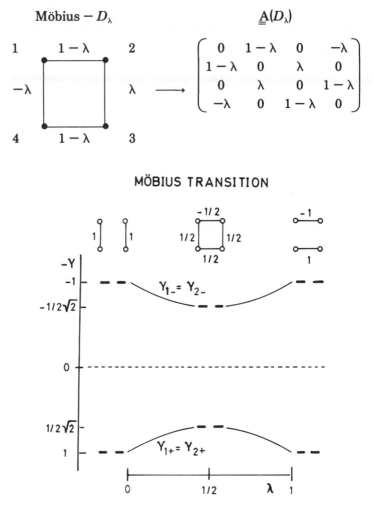

Figure 3.17. Eigenvalue correlation diagram of pericyclic four-center reactions via a Möbius-type transition state.

On solving the corresponding eigenvalue equation (2), one obtains a crossing-free eigenvalue correlation diagram (Figure 3.17). The non-occurrence of a crossing is indicative of a thermally allowed reaction via an aromatic transition situation.

As a generalization of these results, in the case of pericyclic reactions involving an even number of atomic centers which are considered to occur via either Hückel or Möbius transition complexes represented respectively by Hückel and Möbius graphs (44), a closed analytical formula for all eigenvalue correlation diagrams can be derived (2, 22, 31).

The essential features of these equations can be represented in terms of graph-theoretical/topological selection rules for pericyclic reactions (see Table 3.2). By using these rules the results are equivalent to that of the well-known Dewar–Zimmermann and Woodward–Hoffmann rules (11, 42, 45). Note however that their derivation does not require any physical concept, such as orbital considerations or the geometrical structure of the molecular system, but is based only on a logical analysis of the reaction structure and the algebraic formalism of the (single-parameter) λ-model.

Table 3.2. Graph-theoretical/topological selection rules for pericyclic reactions involving n atomic centers (n = even, $\geqslant$ 4)

		n = 4m centers	n = 4m + 2 centers
graph-theoretical model	/2, 22, 31/		
$K_\lambda(^{H}D)$		crossing	no crossing
$K_\lambda(^{M}D)$		no crossing	crossing
Dewar-Zimmermann-rules	/11, 42, 45/		
Hückel		antiaromatic	aromatic
Möbius (anti-Hückel)		aromatic	antiaromatic
Woodward-Hoffmann-rules	/41, 46/		
suprafacial		symmetry forbidden photochemical	symmetry allowed thermal
antarafacial		symmetry allowed thermal	symmetry forbidden photochemical

A comparison between the different models (Table 3.2) shows that thermally forbidden pericyclic reactions are always characterized by a crossing of two eigenvalue correlation lines in the center of the spectrum at $\gamma = 1/2$. These "crossing" eigenvalue lines $\gamma_i(\lambda)$ and $\gamma_j(\lambda)$ have extremum character, i.e. they correspond respectively to the least positive and the largest negative eigenvalue.* As shown below, they play an essential role in the stability analysis of a graph-theoretically described molecular system.

The absolute difference $\Delta\gamma(\lambda) = \gamma_j(\lambda) - \gamma_i(\lambda)$ of these extremum eigenvalue correlation lines can be regarded as a tolerance measure for a pericyclic reaction: the larger the $\Delta\gamma$-values over the whole λ-interval the smoother the thermal reaction should be, whereas $\Delta\gamma = 0$ for some λ indicates a strictly forbidden pericyclic reaction. Employing the expression $\Delta\gamma(\lambda)$, the graph-theoretical/topological selection rules for pericyclic reactions may be stated as follows: Transitions with a crossing-free correlation diagram show a maximum $\Delta\gamma$-value at $\lambda = 1/2$ (aromatic transition situation), while all transitions with crossing correlation lines are characterized by a minimum $\Delta\gamma$-value at $\lambda = 1/2$ ($\Delta\gamma = 0$, antiaromatic transition situation).

3.6.2 The multi-dimensional λ-model

The single-parameter λ-model is now extended to a parametric description of complex reactions with an arbitrary number of reaction parameters. Let $p(\geq 3)$ be the number of reaction partners (reactants, products or intermediates); the reaction lattice is then isomorphic to the lattice $P(p + 1) = 2^{p+1}$ with a diagram of a higher dimensional cube (6, 32). Accordingly, the dynamic sublattice is isomorphic to $P(p) = 2^p$ and thus contains at least one element of the non-mechanistic dimension Δ^2 (see Ch. "Generalized reaction lattice"). Consequently, the choice of the reaction path is no longer unique – in contrast to the single-parameter λ-model for pericyclic reactions with a well defined reaction path (via an aromatic or antiaromatic transition state.). The formal algebraic description of

---

*Remark: In the terminology of HMO-theory, $\gamma_i(\lambda)$ and $\gamma_j(\lambda)$ correspond to HOMO and LUMO energy levels.

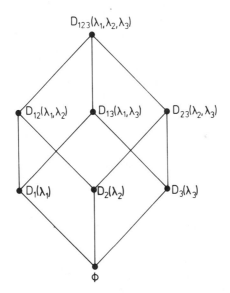

Figure 3.18. Normed, three-dimensional dynamic reaction lattice of a complex reaction.

complex reactions involves an arbitrary number of reaction partners (multi-dimensional λ-model) and is derived in a similar way to that for the single parameter λ-model. Starting from the normalized dynamic sublattice of dimension p (see e.g. Figure 3.18 for $p = 3$), a p-dimensional parameter space $(\lambda_1, \lambda_2, \ldots, \lambda_p)$ is introduced which is bounded by the dynamic graphs $D_k (k = 1, 2, \ldots, p)$. On this basis the dynamic aspect of the respection reactive is described by the set of all graphs $D(\lambda_1, \lambda_2, \ldots, \lambda_p)$ with the constraint that the parameters $\lambda_1, \lambda_2, \ldots, \lambda_p$ fulfill the normalization or simplex condition (31).

$$\sum_{k=1}^{p} \lambda_k = 1; \qquad \lambda_1, \lambda_2, \ldots, \lambda_p \geqq 0; \tag{3}$$

In other words, the set of all dynamic graphs D_k spans a regular Euclidean $(p - 1)$ simplex (47) with respect to variation of parameters $\lambda_1, \lambda_2, \ldots, \lambda_p$ within the given boundaries. This is the domain of definition for all possible reaction states and is termed for short the "reaction domain" D_R. The reaction domain can be conceived of as the system of all possible linear "structural mixtures" of the reaction describing dynamic graphs. In a given mixture characterized by $(\lambda_1, \lambda_2, \ldots, \lambda_p)$ the dynamic edges ("reacting bonds") of the dynamic graph D_k are contained by the fraction λ_k.

In the case $p = 3$ one obtains an equilateral triangle, the vertices of which correspond to the dynamic atoms $D_1(\lambda_1 = 1)$, $D_2(\lambda_2 = 1)$ and $D_3(\lambda_3 = 1)$ (Figure 3.19).* All three line segments, $D_{12}(\lambda_1, \lambda_2)$, $D_{13}(\lambda_1, \lambda_3)$ and $D_{23}(\lambda_2, \lambda_3)$, describe reactions between exactly two reaction partners (reactants and products, reactants and intermediates or products and intermediates), and can be interpreted mechanistically by employing the single parameter λ-model ($\lambda_i + \lambda_j = 1$). Any internal point of the triangle depicts a non-empty mixture of all three graphs D_k and represents thus a "non-mechanistic" situation; in other words, any reaction path between D_i and D_j involving at least one internal triangular point cannot be interpreted mechanistically in the frame of classical structure theory. Consequently, an infinite number of possible reaction paths exist, none of them being distinguished; that is, each of them can be realized in a particular experiment. On the other hand, if one introduces additional physical or chemical criteria, a most probable reaction path can be selected based upon a statistical interpretation.

---

*Remark: A similar coordinate system, the axes of which are the sides of an equilateral triangle, is used in thermodynamics to describe compound mixtures consisting of three components (48). In this case, the vertices represent pure components, the line segments correspond to binary mixtures, and every internal point of the triangle describes a certain composition of the ternary system. The molar fractions of the components in such a compound mixture correlate with the λ_k–parameter of dynamic graphs in a "structural mixture". The set of all possible component distributions is the domain of definition for the respective (triangular) phase diagram.

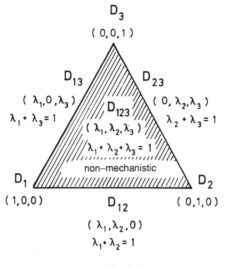

Figure 3.19. Simplex representation (reaction domain D_R) of the normed three-dimensional dynamic sublattice.

For this purpose we consider the matrix function

$$\underline{\underline{A}}[D(\lambda_1, \lambda_2, \ldots, \lambda_p)] = \sum_{k=1}^{p} \lambda_k \underline{\underline{A}}(D_k) \qquad (4)$$

constrained by the simplex condition (3). Solution of the corresponding eigenvalue problem yields $(p-1)$ dimensional hypersurfaces over the simplicial domain D_R of definition. In the same way as for the single-parameter λ-model, the absolute difference

$$\Delta\gamma(\lambda_1, \lambda_2, \ldots, \lambda_p) = |\gamma_j(\lambda_1, \lambda_2, \ldots, \lambda_p) - \gamma_i(\lambda_1, \lambda_2, \ldots, \lambda_p)| \qquad (5)$$

of the correlation hypersurfaces, corresponding respectively to the least positive or most negative eigenvalue for any structural mixture ("HOMO-LUMO" differences in HMO-theory) shall be regarded as a tolerance measure for the reaction. If $p = 3$, $\Delta\gamma(\lambda_1, \lambda_2, \lambda_3)$ can be represented by contour lines of constant value over the triangular reaction domain.

Let us now define a single reaction path $p(\lambda) = p[\Delta\gamma(\lambda)]$ on the $\Delta\gamma$-hypersurface characterizing a transition between two reaction partners. For this purpose it is necessary to introduce $(p - 2)$ additional hypersurfaces $f_l(\lambda_1, \lambda_2, \ldots, \lambda_p)$ $(l = 1, 2, \ldots, p - 2)$ of dimension $(p - 1)$ over D_R such that their intersection yields a projection of the reaction domain onto a single parameter curve in D_R (variational parameter: λ) connecting the respective dynamic graphs D_i and D_j. The functions f_l may be correlated with the "non-energetic" aspects of chemical reactions, such as the polarity of substituents or solvent effects.

The length l_p of a reaction path between two reaction partners i and j can be defined as the line integral of $p[\Delta\gamma(\lambda)]$ in the interval $\lambda = 0$ $(D_i) \leq \lambda \leq \lambda = 1$ (D_j) (31). On this basis a (non-energetic) condition for the most favorable reaction path can be formulated:

"The shorter the reaction path, i.e. the smaller l_p, the more likely the reaction is to occur via this route".

The claim for the minimum length of a reaction path corresponds to Butlerov's principle of minimal structural change during a chemical reaction (22, 49) as well as to Ugi's principle of minimal chemical distance (50).

As an example we consider the (previously discussed) polar [2 + 2] cycloaddition involving a zwitterionic intermediate (see Figure 3.9) (30). The adjacency matrices, $\underline{\underline{A}}(D_1)$ and $\underline{\underline{A}}(D_2)$, of the reactants and products are those of a single parameter pericyclic transition (see above), whereas the adjacency matrix, $\underline{\underline{A}}(D_3)$, of the intermediate is given by:

$$\underline{\underline{A}}(D_3) = \begin{pmatrix} 0 & 0 & 0 & 0 \\ 0 & 0 & 1 & 0 \\ 0 & 1 & 0 & 0 \\ 0 & 0 & 0 & 2 \end{pmatrix}$$

or, in relation to the center of the spectrum,

$$\underline{\underline{A}}(D_3) = \begin{pmatrix} -1 & 0 & 0 & 0 \\ 0 & 0 & 1 & 0 \\ 0 & 1 & 0 & 0 \\ 0 & 0 & 0 & +1 \end{pmatrix}$$

The resulting contour diagram of constant $\Delta\gamma$-values is depicted in Figure 3.20. It is worthwhile mentioning that in the case of cyclo-additions considered to occur via diradical intermediates the same contour diagram is obtained by employing the formalism of the multi-dimensional λ-model (31). The shape of the contour diagram leads to the conclusion that thermally forbidden pericyclic reactions may be converted into (more or less) allowed processes with par-ticipation of the discussed intermediates. In other words, any non-mechanistic reaction path between D_1 and D_2 under the influence of D_3 is posisible and should be observed in a particular experi-ment. Similarly, one may draw the conclusion that a stepwise mechanism involving isolable intermediates occurs under suitable conditions.

On the other hand, owing to the statistical interpretation of non-mechanistic reaction paths, only probabilistic statements concerning the result of a single experiment are permitted. To demonstrate this in more detail we have calculated by numerical integration the length L_p of four different reaction paths between D_1 and D_2 (see Figure 3.20). According to the principle of the shortest reaction path, one should expect that the most probable course of reaction is path (c) and that path (d) is the least likely. From this result we draw the conclusion that on the one hand participation of zwitterion or diradical intermediate structures is necessary in order to enable [2 + 2] cycloadditions to occur by avoid-ing the crossing point, but on the other hand the formation of isolated intermediates will be very unlikely. This result corresponds to the experimental observation that [2 + 2] cycloadditions occur more smoothly if the reaction participants bear major polarizing substituents or if the reaction takes place in a strong polar solvent (30). We mention here that zwitterion or diradical intermediates can be detected only under extreme reaction conditions. Taking into account the additional supposition that larger maximum $\Delta\gamma$-values (or a more avoided crossing) indicate a low activation energy, the reaction path (c) should also be energetically favoured over (e.g.) path (b), as accords with experimentally observed substituent and solvent effects (30).

Let us conclude our exposition of the multi-dimensional λ-model by mentioning a hitherto neglected problem. The described for-malism is implicitly based upon the assumption that the dynamic

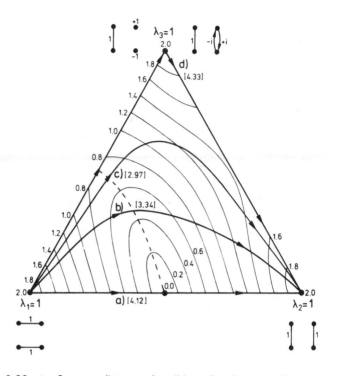

Figure 3.20. $\Delta\gamma$-Contour diagram describing $(2 + 2)$-cycloadditions involving zwitterion or diradical intermediates.

graphs D_k spanning the reaction domain D_R are pairwise disjoint. This condition, however, is in general fulfilled only if the graphs D_k are conceived of as abstract points, i.e. only if their attribute as o-simplices spanning D_R is considered. A more specific interpretation reveals that the graphs D_k are generally not disjoint with respect to intersections of their edge sets, as can be seen in the case of polar $[2 + 2]$ cycloaddition. We are thus confronted with the same difficulty as discussed previously in connexion with the abstract and concrete interpretation of multi-dimensional Boolean reaction lattices.

Hence, the multi-dimensional λ-model will be valid in the strict sense only if the edge sets of the graphs D_k are pairwise disjoint (i.e. no mesomeric structure is partially contained in another). In this case the whole system is uniquely described by adding the edge sets of D_k according to eq. (4). Applying the terminology of measure theory, one can say that the edge sets are σ-additive for all structural mixtures (51).

In the general case of the multi-dimensional λ-model, σ-additivity does not hold. We presume, however, that the simplex condition guarantees a weaker kind of additivity, even if some of the superimposed graphs have a non-empty intersection of their edge sets. We also do not rule out the possibility that a non-Boolean probability theory based upon orthomodular lattices (52) should be applied in order to understand these problems without any contradiction*.

3.5.1.3 *Consideration of constant influences on the reaction*

Another extension of the single parameter λ-model (and analogously of the multi-dimensional λ-model) is the inclusion of influences that remain invariant during the reaction. According to the analysis of the three-dimensional reaction lattice, such invariant perturbations may be correlated with the structure of the static graph. Depending on the nature of S we distinguish three essential classes of constant influences:

(i) vertex perturbations or "α-perturbations" (for example substituent effects) (2). i.e. the superposition of D and a totally disconnected pseudograph S. This implies that additional weighted loops are introduced on the edge set of D;

(ii) edge perturbations or "β-perturbations" (through bond or through space interactions) (2), i.e. the introduction of additional weighted edges to the vertex set of D;

(iii) graph perturbations or "γ-perturbations" (for instance, catalytic effects) (60), i.e. the superposition of D by another graph K with the following properties:

   – The intersection of the vertex sets of K and D is non-empty

---

*The problems outlined above are discussed in more mathematical detail in papers (6, 53).

or – in other words – K contains at least one additional vertex compared to D.

– At least one edge of K is incident with such a vertex.

In all cases the λ-dependent matrix function (1) (single parameter λ-model) is extended by including a "static term" $\underline{A}(^gS)$ (2):

$$\underline{A}(g, \lambda) = \underline{A}(gS) + (1 - \lambda)\,\underline{A}(D_1) + \lambda\underline{A}(D_2) \qquad (6)$$

with ^gS belonging to one of the structural classes (i), (ii), or (iii) and g equal to the weight of its influence on the transition. If we always conceive of perturbations as positively weighted (i.e. only the sign of the edges of S, but not that of the weight of a perturbation may change), g will vary within the interval $0 \leq g < \infty$. As a consequence, the domain of definition of all (g, λ)-dependent situations is a band of width 1 and infinite length, which is bounded on three sides by $\lambda = 0$, $\lambda = 1$ and $g = 0$ (see Figure 3.21).

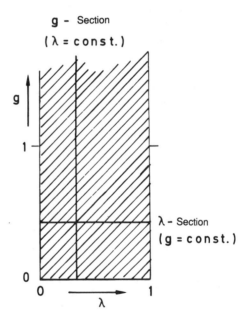

Figure 3.21. Domain of definition of the single parameter λ-model involving constant influences on the transition.

In order to get information about the essential features of the g- and λ-dependent correlation diagrams of eigenvalues γ_i, obtained by solving the eigenvalue problem of the relevant matrix function (6), we consider two orthogonal projections of the reaction domain:

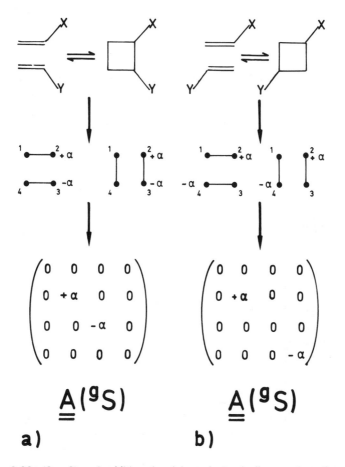

Figure 3.22. (2 + 2)-cycloaddition involving electronically counteracting substituents. (a) *cis*-arrangement (b) *trans*-arrangement ($X =$ electron-releasing substituent $Y =$ electron-withdrawing substituent).

a) λ-variation in the interval $0 \leqq \lambda \leqq 1$ at constant g ($\neq 0$, e.g. $g = 0.5$)

b) g-variation in the interval $0 \leqq g < \infty$ by keeping λ invariant (e.g. $\lambda = 0.5$).

For example, we may reconsider the previously mentioned (2 + 2) cycloaddition between the ethylene derivatives with electronically counteracting substituents (i.e. one molecule bearing an

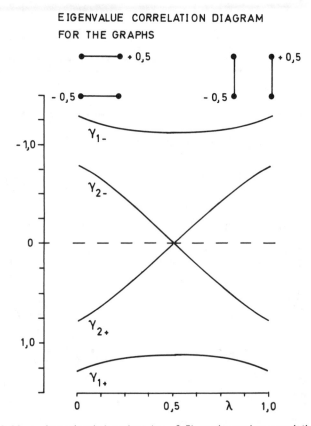

Figure 3.23. γ-dependent/g-invariant ($g = 0.5$) γ_r-eigenvalue correlation diagram of cycloadditions involving *trans*-arranged substituents with counteracting electronic behavior.

electron-releasing group (e.g. $-NMe_2$) and the other an electron-withdrawing group (e.g. $-CO_2Me$)) (see Figure 3.9). Two different arrangements of the substituents are possible with respect to the pericyclic transition complex: a) *cis*- and b) *trans*-positioning (Figure 3.22). For simplicity, we shall assume that the effects of the substituents are of the same magnitude (but of opposite direction).

Let $+\alpha(\geqq 0)$ be the influence of the electron-releasing substituent on the respective C-atom, and $-\alpha$ that of the electron-

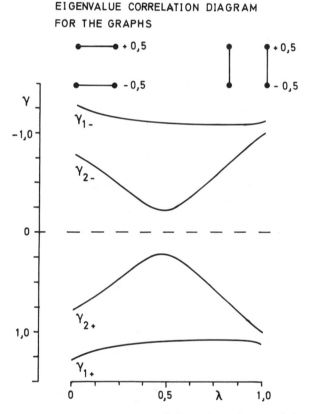

Figure 3.24. γ-dependent/g-invariant ($g = 0.5$) γ_i-eigenvalue correlation diagram of cycloadditions involving *cis*-arranged substituents with counteracting electronic behavior.

withdrawing group; then the graphs ^{g}S ($g = \alpha$) depicted in Figure 3.22 will describe polar cycloadditions with *cis-* and *trans-*aligned substituents, respectively. The corresponding matrices $\underline{A}(^{g}S)$ are shown in Figure 3.22 as well. If one solves the eigenvalue problems of the respective matrix functions $\underline{A}(g, \lambda)$ (equations (6) and (2)) at

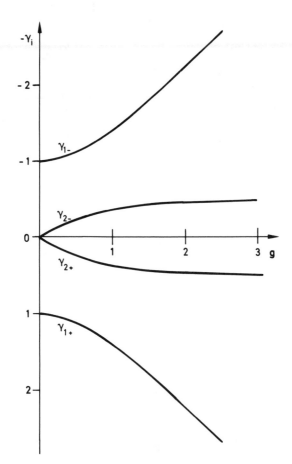

Figure 3.25. *g*-dependent/λ-invariant ($\lambda = 0.5$) γ_i-eigenvalue variation diagram of cycloadditions involving *cis*-arranged substituents with counteracting electronic behavior.

$g = \alpha = 0.5$ in the interval $0 \leq \lambda \leq 1$, one obtains λ-dependent/g-invariant γ_i-eigenvalue correlation diagrams for both transitions (Figures 3.23 and 3.24). Figures 3.25 and 3.26 show the corresponding g-dependent/λ-invariant γ_i-eigenvalue variation diagrams.

It follows from these diagrams that the crossing observed in

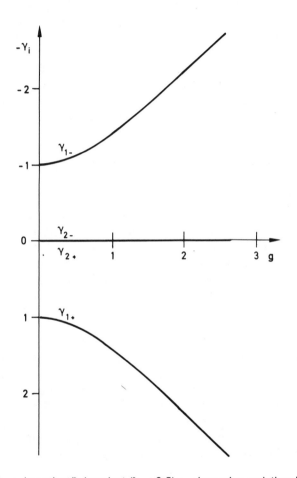

Figure 3.26. *g*-dependent/λ-invariant ($\lambda = 0.5$) γ_i-eigenvalue variation diagram of cycloadditions involving *trans*-arranged substituents with counteracting electronic behavior

the unperturbed pericyclic four-center transition via a Hückelring (Figure 3.16) is avoided in the case of polar cycloadditions involving *cis*-arranged substituents X and Y. Figure 3.25 shows in addition that $\Delta\gamma$ increases with increasing g values and approaches a limit $(\Delta\gamma = 0.5)$ for $g \to \infty$. On the other hand, if the substituents are *trans*-arranged, the crossing is preserved independently of the g value. Consequently, in the first case the thermally forbidden non-polar pericyclic reaction is turned into an allowed process, in particular if the *cis*-aligned substituents are strongly polar; whereas in case of *trans*-aligned, equally counteracting substituents the reaction remains forbidden, irrespective of their strength.

This result is in agreement with experimental observations (30) as well as with the conclusions drawn from the multi-dimensional λ-model.

3.6.4 Generalization of the λ-model for the description of systems involving heteroatoms

In the following, a general formulation of the λ-model is given suitable for treating chemical reactions involving heteroatoms. This generalization is based upon the following assumptions:

(i) The reaction is formally divided into an intrinsic dynamic part (reaction domain D_R) and a part of constant influences ^gS which is extraneous with respect to the reaction itself.

(ii) The discrete structure of heteroatom effects is described by coloring vertices and edges of both the dynamic and static parts.

(iii) The strength of heteroatom effects is modelled by a continuous variation of the "depth" of a given coloring (grey value).

In the general case, the coloring and its grey value is described by two tensors $\underline{F}$ and $\underline{G}$, which are multiplied element-wise respectively, with the matrices of the edge values of S and D_k (54). If one differentiates by respective indexing between "static" and "dynamic" "color tensors" ($\underline{F}_s$ and $\underline{F}_D$) and "grey tensors" ($\underline{G}_s$ and $\underline{G}_D$), one obtains the following matrix function $\underline{\underline{A}}$ $(\lambda_1, \lambda_2, \ldots, \lambda_p)$ over the reaction domain D_R:

$$\underline{\underline{A}}(\lambda_1, \lambda_2, \ldots, \lambda_p) = \underline{G}_s \circ \underline{\underline{F}}_s \circ \underline{\underline{A}}(S) + \underline{G}_D \circ \underline{F}_D \circ \sum_{k=1}^{p} \lambda_k \underline{\underline{A}}(D_k)$$

where
$$\sum_{k=1}^{P} \lambda_k = 1; \ \lambda_k \geq 0; \quad k = 1, 2, \ldots, p \tag{7}$$

$$g_{ij}(\underline{\underline{G}}_s), \quad g_{ij}^{(D)}(\underline{\underline{G}}_D) \in \mathbb{R}_0^+$$

$$f_{ij}(\underline{\underline{F}}_s), \quad f_{ij}^{(D)}(\underline{\underline{F}}_D) \in \mathbb{Z}.$$

The composition " ∘ " introduced above is defined as a inner product of two matrices (tensors) $\underline{\underline{A}}$ and $\underline{\underline{B}}$, where only elements with the same index pair are multiplied thus:

$$a_{ik} \in \underline{\underline{A}}, \ b_{lj} \in \underline{\underline{B}}; \quad \underline{\underline{A}} \circ \underline{\underline{B}} = \underline{\underline{C}} = (c_{ij}) = (a_{ij} \circ b_{ij}). \tag{8}$$

In order to express this composition in terms of tensor operations, we form at first the tensor product $\underline{\underline{A}} \oplus \underline{\underline{B}}$ of $\underline{\underline{A}}$ and $\underline{\underline{B}}$, i.e. every element of $\underline{\underline{A}}$ (first factor) is multiplied with every element of $\underline{\underline{B}}$ (second factor):

$$a_{ik} \oplus b_{lj} = w_{iklj}.$$

Subsequently, a transaction of k to l is carried out, where k is set equal to l:

$$a_{ik} \oplus b_{kj} = w_{ikkj}.$$

The same result is obtained by scalar multiplication of the tensor product w_{iklj} with the product of both the Kronecker symbols δ_{ik} and δ_{lj}:

$$\delta_{ik} \circ \delta_{lj} \circ w_{ikej},$$

or, equivalently, by tensor multiplication of w_{lklj} with the unit tensor δ_{ek} and assuming contraction:

$$\delta_{lk} \oplus w_{iklj} = w_{lklj} = w_{ij}.$$

In order to investigate heteroatom effects on pericyclic reactions or on the behavior of aromatic systems, it is sufficient to consider the dynamic term of eq. (7).

In the case of (1, 3)-diazole the reduced minimum form is given by:

$$
\begin{pmatrix}
\alpha & 1 & 1 & 1 & 1 \\
1 & \alpha & 1 & 1 & 1 \\
1 & 1 & \alpha & 1 & 1 \\
1 & 1 & 1 & \alpha & 1 \\
1 & 1 & 1 & 1 & \alpha
\end{pmatrix}
\circ
\begin{pmatrix}
1 & 1 & 1 & 1 & 1 \\
1 & 0 & 1 & 1 & 1 \\
1 & 1 & 1 & 1 & 1 \\
1 & 1 & 1 & 0 & 1 \\
1 & 1 & 1 & 1 & 0
\end{pmatrix}
\circ
\begin{pmatrix}
1 & 1 & 0 & 0 & 1 \\
1 & 1 & 1 & 0 & 0 \\
0 & 1 & 1 & 1 & 0 \\
0 & 0 & 1 & 0 & 1 \\
1 & 0 & 0 & 1 & 0
\end{pmatrix}
=
\begin{pmatrix}
\alpha & 1 & 0 & 0 & 1 \\
1 & 0 & 1 & 0 & 0 \\
0 & 1 & \alpha & 1 & 0 \\
0 & 0 & 1 & 0 & 1 \\
1 & 0 & 0 & 1 & 0
\end{pmatrix}
$$

or, in relation to the center of the matrix, we obtain:

$$
\begin{pmatrix}
3/5\alpha & 1 & 0 & 0 & 1 \\
1 & -2/5\alpha & 1 & 0 & 0 \\
0 & 1 & 3/5\alpha & 1 & 0 \\
0 & 0 & 1 & -2/5\alpha & 1 \\
1 & 0 & 0 & 1 & -2/5\alpha
\end{pmatrix}.
$$

The "grey value" α can vary within the interval $0 \le \alpha < \infty$.

Starting from the Kekulé formulas for (1, 3)-diazole (see Figure 3.27), this reduced form results if the respective dynamic graph, constructed as shown in Figure 10, is multiplied with a coloring tensor determined by the two N-atoms as well as by a grey tensor with equal grey values α in the diagonal. These grey values affect only vertices having a "non-zero-color", i.e. they correspond to N-atoms[+].

3.7 Stability Analysis of Systems Described Graph-theoretically

The chemist's concept of aromaticity is strongly connected with the idea of a particular stable state of low energy.

In analogy to Hückel MO-theory, the eigenvalues of dynamical graphs D are in general interpreted energetically by making use of graph-theoretical structure theory. This gives us the possibility of speaking about the stability of a system described by graphs in energetic terms.

[+]Note that the "color" of a carbon atom is set to zero.

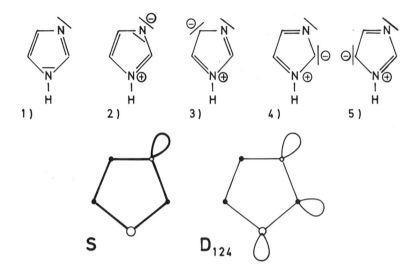

Figure 3.27. Mesomeric structures of the (1, 3)-diazole, the set of correlated graphs: S (static graph), and D_{124} (reduced, minimal, dynamic subgraph of the dynamic graph D).

A molecular realization of a graph is "stable" only if the eigenvalues are realized by corresponding states of electrons, for which we have $\gamma_i > 0$ (or $\varepsilon_i < 0$ energy eigenvalue).

The use of the concept of stability is plausible in this connection, and there is nothing to contradict it, but its introduction in our field of interest may seem strange since it was not deduced from stability analysis of the molecular system described by graphs.

As Wiedemann has shown, it is possible in principle to understand the dynamic graphs as a dynamic system. For this reason one can subject the graphs to a stability analysis, and as a consequence one can substantiate the intuitive introduction of the idea of stability into the graph-theoretical concept of aromaticity (5).

The description of pericyclic reactions by the dynamic graph D suggests connecting the graph D with the time development of the system during reaction.

By viewing the reaction as a perturbation of the molecular system, the time development of the reaction may be given by the structure of the graph D. We express this by taking the adjacency matrix A_D of D as an operator affecting the perturbation determining the time development. For instance, let us assume that

$$-\frac{dX}{dt} = A_D \cdot X,$$

where X describes the perturbation at time t, and that:

$$X = C \cdot e^{-\Gamma t}.$$

Following our premises, the perturbation X, taken again as the "situation of the reaction" depends on the underlying structure of the reaction.

Therefore there may be the following connection between A and C;

$$AC = \Gamma \cdot C$$

where C is the matrix of the eigenvectors and Γ the matrix of the eigenvalues of A. This expression corresponds to the eigenvalue problem of A written in matrix form, which is independent of time (time-free). Now we can understand the perturbation X to be a system C of eigenvectors that can be contracted or expanded by the time factor $e^{-\Gamma t}$ corresponding to Γ. Since $X = C$ holds for $t = 0$, one concludes that the perturbation at time $t = 0$ can be totally represented by the set of components of the eigenvectors of the system C. In this way the real existence of a system creates its own perturbation.

At best this means for positive eigenvalues that the system is asymptotically stable since time $t > 0$ is always increasing.

As a consequence of the proposal made here, one can speak only about the structure of a system and its stability if time is understood as an external parameter of the expanding matrix $e^{-\Gamma t}$.

Formal expression of this statement is obtained from solution of the connected differential equation, provided the solution is based upon the proposal mentioned above:

Conditions: $A \cdot C = C \cdot \Gamma$

$\qquad\qquad\quad A = $ Hermitean matrix

C = eigenvector matrix

Γ = Diag. (γ_j) = eigenvalue matrix

Def.: $e^{-\Gamma t} = \text{Diag.} \left(e^{-\gamma_j \cdot t}\right)$

Assumption: $-\dfrac{dX}{dt} = A \cdot X$

Proposal: $X = C \cdot e^{-\Gamma t}$

If one inserts the proposed solution into the differential equation one gets:

$$-\frac{dX}{dt} = -\frac{d}{dt}\left(Ce^{-\Gamma t}\right)$$

$$= -C\frac{d}{dt}\left(e^{-\Gamma t}\right)$$

$$= -C\frac{d}{dt}\left\{\text{Diag}\left(e^{-\gamma_j t}\right)\right\}$$

$$= -C\,\text{Diag}\left\{-\gamma_j e^{-\gamma_j t}\right\}$$

$$= C \cdot \Gamma e^{-\Gamma t}$$

Moreover, corresponding to our solution, the equation

$$AX = ACe^{-\Gamma t}$$

holds, and, including the eigenvalue equation, it follows that:

$$ACe^{-\Gamma t} = C\Gamma e^{-\Gamma t}.$$

The assumption that $-\dfrac{dX}{dt} = AX$ is therefore now established.

If one attempts to understand the expansion factor as an inner parameter, taking the expansion matrix as an operator acting on the structure C, the differential equation can be solved only in the case of totally unstructured and unconnected graphs.

Making a corresponding attempt at solution:

$$X = e^{-\Gamma t}C$$

the differential equation assumes the form:

$$-\frac{dX}{dt} = \frac{d}{dt}\left(e^{-\Gamma t}C\right)$$

$$= \Gamma e^{-\Gamma t}C$$

From our trial solution, it is seen that the following equation holds:

$$AX = Ae^{-\Gamma t}C$$

So the assumption that $-\dfrac{dX}{dt} = AX$ can be fulfilled only if $A = \Gamma$ is a diagonal matrix. Otherwise the components of the eigenvalues of C are mixing. This means that the eigenvalue equation cannot be satisfied or equivalently that we cannot speak of a structure of the system in such a case.

The general statement that the system possesses a structure is followed by the fact that one can observe the system at different times. So time itself becomes an external parameter with respect to the system under consideration.

The matrix form in which the statements are expressed leads to the conclusion that one can describe the system in its totality in terms of its structure. For this reason, the statement on the stability of graphs refers to the set of all eigenvalues λ_i. The graph A is stable if its spectrum is positive or non-negative (55–57).

In all cases, connected simple graphs without loops possess negative eigenvalues too. A system described by such a graph is neccessarily unstable with respect to its totality.

However, if we wish to speak about stability in such a case we must introduce a second condition enabling us to observe subsets of the set of all states on its own. A measure of aromaticity could be based on the consideration of such subsets.

For this purpose we now divide the eigenvector space into two subspaces spanned by the vectors belonging to the negative and to the non-negative (or positive) eigenvalues.

With the additional postulate that of any two eigenvalues one will be realized preferentially since its perturbation decays faster according to the term $c_i \cdot e^{-\gamma_i t'}$, we shall designate the realized eigenvalue as the more stable eigenvalue. Assuming that the realization is a stochastic process described by matrix A, this postulate means that the more stable of the two eigenvalues is realized more frequently by random events.

Moreover, let us suppose the system to be ergodic, i.e. in the course of time all its accessible states will be realized. With a given probability the system will occupy even states, though these are unstable because of their negative eigenvalues. The probability of occupying the state corresponding to the eigenvalue γ_i may be proportional to $e^{-\gamma_i t}$ for any time t. Then the conditional probability for the transition from the state with eigenvalue γ_i to the state with eigenvalue γ_j is proportional to the expression:

$$\frac{e^{-\gamma_i t}}{e^{-\gamma_j t}} = e^{-(\gamma_i - \gamma_j)t}$$

Now, if we seek the maximal probability for transition of the system from a stable ($\gamma_i > 0$) to an unstable state ($\gamma_j < 0$), one is asking for the minimal difference ($\gamma_i - \gamma_j$) where γ_i belongs to the non-negative half-space and γ_j to the negative one. Thus, the difference between the minimal positive and the minimal negative eigenvalues of the graphs describing the system is decisive for its stability.

3.7.1 Stability Analysis of Cyclic Graphs

Let us consider the spectrum of the dynamic graphs for a four-membered pericyclic reaction depending on the reaction parameter λ ($0 \leqslant \lambda \leqslant 1$). When $\lambda = 0.5$ the lowest positive and the lowest negative eigenvalues come together at the zero-line ($\gamma_{+,\min} = \gamma_{-,\min} = 0$). Moreover, for the eigenvalue functions $\gamma_{+,\min} = f_+(\lambda)$ and $\gamma_{-,\min} = f_-(\lambda)$ the derivative $d\gamma/d\lambda$ shows a point of discontinuity for this argument.

Colloquially, it may be said that the functions of both eigenvalues do not cross each other at this point. On the other hand, the eigenvalue functions of chemical reactions represented by a dynamic sublattice that show similar behavior are "thermally forbidden" (2, 22).

If we wish to stay with our interpretation of graph-theoretical structure theory, we need to derive the forbidding of the reaction in terms of the "crossing" of the eigenvalue functions $\gamma = \gamma(\lambda)$ at the zero-line $\gamma = 0$ for $\lambda = 0.5$.

Now, we considered our system to be ergodic with respect to the realization of eigenvalues for any $\lambda \in [0, 1]$. From this we concluded that the difference of the lowest positive and lowest negative eigenvalue governs the stability of the system.

For $\lambda \to 0.5$ the difference tends to zero and the probability of the system to change into an unstable stage tends to one, i.e. it tends to certainty. Simultaneously the unstable state becomes less unstable. For $\lambda = 0.5$ this state loses its unstable character. The eigenvalue $\gamma = 0$ means that the system is in a state characterized by just this eigenvalue. It will be stable in the sense that any perturbation that occurs at time t_0 is neither growing nor diminishing with respect to time. The system is stable but not asymptotically stable.

3.8 Time Independence Distribution of the Perturbation of the Eigenvalues

Let us assume that the system is permanently exposed to perturbations which are conditional on its own structure. The time-independent distribution of the fluctuations with respect to the eigenvalue γ_i may be given by the distribution function

$$\gamma_i^2 e^{-\gamma_i^2 (\Delta\gamma)^2}$$

where $(\Delta\gamma)$ represents the diviation of the eigenvalue perturbed by fluctuation from the eigenvalue arising from the structure of the unperturbed graph.

The realization of the eigenvalues now requires us to normalize the expression:

$$\int_{-\infty}^{+\infty} \gamma^2 e^{-\gamma^2 (\Delta\gamma)^2} d(\Delta\gamma) \overset{!}{=} 1$$

In this way one can interpret $\gamma^2 \cdot e^{-\gamma^2 (\Delta\gamma)^2}$ as a density function.

The complete function, representing the distribution of the eigenvalues γ_i about the relative position of the center of the function to the origin, now becomes

$$\gamma_i^2 e^{-\gamma_i^2 \left[(\Delta\gamma)^2 + \gamma_i^2 \right]}$$

Figure 3.28 represents schematically the eigenvalue functions $\gamma_{+, \min} = f_+(\lambda)$ and $\gamma_{-, \min} = f_-(\lambda)$ and eigenvalues of the system within the reach of fluctuations.

For large absolute values of eigenvalues the deviation from the given eigenvalue γ_i becomes very small and tends to zero for $\gamma_i \to \infty$. The result is the Dirac σ function (see Figure 3.29).

For small absolute values of eigenvalues γ_i fluctuations with large amplitudes become much more probable; so, for $\gamma_i = 0$ any

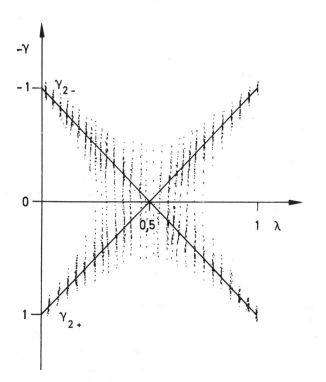

Figure 3.28. Illustration of two eigenvalue functions $\gamma_i = \gamma(\lambda)$ crossing on the zero line, and the corresponding events realizing these eigenvalues by fluctuation. It was assumed that the frequency of events is always equal for each value of λ.

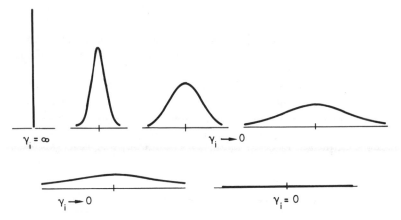

Figure 3.29. Schematic representation of the normalized distribution of perturbated eigenvalues for different expectation values $\overline{\gamma}$.

fluctuation is of equal frequency. Even fluctuations with "infinitely large" amplitudes are allowed.

The approach of the eigenvalue function $\gamma_{+,\,\min} = f_+(\lambda)$ to the zero line leads to fluctuations of the system becoming ever larger. Along such lines, states can be realized in which the system is unstable. Hence, for $\gamma_i = 0$ states can be realized from which the system decomposes instantaneously since $\Delta\gamma \rightarrow \infty$.

In the sense of this procedure, the crossing of eigenvalue functions at the zero line means that any arbitrarily large fluctuations are of equal probability. In the case that both eigenvalue functions $\gamma_{+,\,\min} = f_+(\lambda)$ and $\gamma_{-,\,\min} = f_-(\lambda)$ do not cross each other on the zero line, the situation is comparable to the former one if they closely approach each other.

Only, when both eigenvalue functions cross each other sufficiently far from the zero line may the corresponding chemical system be able to follow the path described by λ.

The systems of interest are characterized by the fact that states are realized in which the eigenvalues of the dynamic graphs are placed or cross on the zero line. The description mentioned above of such systems enables us to understand on a graph-theoretical basis ideas like the automerism of valencies, and bond fluctuation in antiaromatic molecules.

If realized states are located on the zero line or if they cross each other on this line, the behavior of the corresponding molecular system should established by the necessary fluctuation.

But the fact that the system is in the process of realizing such a state is highly improbable.

However, it is just this situation which is expressed by the term "tautomerism" of valencies. Depending upon the influence of the static graph on the matrix polynomial, one is able to distinguish between isomerism and tautomerism or bond fluctuation and mesomerism, respectively.

As can be seen, both of the latter ideas are related in systems without any crossing of the eigenvalue functions on the zero line.

Summary

Using benzene-like aromatic systems and pericyclic reactions with an even number of centers, the principles of graph-theoretical structure theory are described and extended to conjugated heterocycles and cyclic systems with an odd number of centres. With topological analysis of the graphs of these systems as a foundation, a graph-theoretical definition of the idea of aromaticity in regard to monocyclic compounds is presented.

Moreover, it can be shown that in the case of monocyclic systems the graphs belonging to pericyclic reactions with an even number of centres form a Boolean lattice with three atoms.

Contrary to the above, the systems of graphs belonging to condensed aromatic species and conjugated monocyclic systems having an odd number of centres do not form a lattice but a semilattice.

In general, an abstract Boolean diagram with a dimension greater than three underlies these semilattices. The reaction lattices as well as the reaction semilattices contain dynamic sublattices or semilattices. The abstract diagrams on which they are based allow the definition of a norm to describe the reaction path continuously by eigenvalue correlation diagrams. The normed lattice can be interpreted statistically. The λ-model or the multi-λ-model for two and higher dimensional dynamic sublattices is based on this procedure. This gives us the possibility of developing selection rules to decide if a reaction is allowed or not. In the case of mechanistically explicable pericyclic reactions having two-dimensional Boolean sublattices, the above mentioned rules are comparable to the Woodward–Hoffmann rules.

Since the normed dynamical lattices describe the dynamic behavior of the system, the related edge weighted graph can be analyzed using Ljapunov stability analysis.

This leads to a graph-theoretical foundation for the non-crossing rule of eigenvalue functions on the zero-line related to the forbidding of a reaction. Using a time-independent description of the distribution of the perturbed eigenvalues, an interpretation based on probability theory permits statements on the appearance of valence isomers in antiaromatic systems to be made.

3.9 References

1. I. Ugi, D. Marquarding, H. Klusacek, G. Gokel, P. Gillespie; Angew. Chem. 82, 18 (1970) 741.
2. E.C. Hass, P.J. Plath; Z. Chem. 22, 1 (1982) 14.
3. R.F.W. Bader; "A Theory of Molecular Structure"; Studies Phys. Theor. Chem. 28 (1983) 40.
4. P.G. Mezey; Stud. Phys. Theoret. Chem. 28 (1983), 75.
5. M. Wiedemann; "Graphentheoretische Behandlung des Aromatizitätsbegriffes"; Lecture, Pädag. Hochsch. Köthen (1983).
6. P.J. Plath; "Logik als Struktur chemischer Reaktionen – ein Programm zur Analyse der chemischen Formelsprache"; special printing (Bremen) and: "Diskrete Physik molekularer Umlagerungen". Teubner Texte zur Physik, Bd. 19, BSB B.G. Teubner Verlagsgesellschaft, Leipzig (1988).
7. H. Diegruber; "Der Modellcharakter des Begriffes Aromatizität", Staatsexamensarbeit, University of Bremen (1977).
8. H. Primas; "Chemistry, Quantum Mechanics and Reductionism" in "Lecture Notes in Chemistry", Springer Verlag, Berlin, Heidelberg, New York (1981).

9. R.I.G. Hughes; Spektr. Wissen. Heft 12 (1981) 85.
10. H.H. Günthard, H. Primas; Helv. Chim. Acta 39, 6 (1956) 1645.
11. M.J.S. Dewar; Angew. Chem. 83, 22 (1971) 859.
12. M.J. Goldstein, R. Hoffmann; J. Amer. Chem. Soc. 93, 23 (1971) 6193.
13. A. Graovac, I. Gutman, N. Trinajstić, T. Živković; Theor. Chim. Acta 26, (1972) 67.
14. I. Gutman, N. Trinajstić; Top. Curr. Chem. 42, (1973) 49.
15. Reports and Reviews of the Microsymposium: "Graph Theory in Chemistry"; Match 1, Hrsg. A.T. Balaban, A. Dreiding, A. Kerber, O.E. Polansky; Mülheim/Ruhr (1975).
16. A.T. Balaban, D. Farcasiu, R. Banica; Rev. Roum. Chim. 11, (1966) 1205.
17. C. Berge; "Introduction à la Théorie des Hypergraphes", Les Presses de l'Université de Montreal, (1973).
18. E.C. Hass, P.J. Plath; Match 1, (1975) 141.
19. U. Döring, P.J. Plath; Match 1, (1975) 151.
20. E.C. Hass, P.J. Plath; Brem. Brief Chem. 1, 1 (1977) 6.
21. P.J. Plath, E.C. Hass; Croat. Chim Acta 51, 3 (1978) 225.
22. E.C. Hass, P.J. Plath; Brem. Brief. Chem. 2, 2 (1978) 3.
23. K.J. Laidler; "Reaktionskinetik I", BI - Hochschultaschenbücher vol. 280, Bibliographisches Institut Mannheim, Vienna, Zürich (1970) 74.
24. P.J. Plath; Match 7 (1979) 229.
25. A.T. Balaban; Rev. Roum. Chim 12, (1967) 975.
26. J.B. Hendrickson; Angew. Chem. 86, 2 (1974) 71.
27. G. Wildeboer, P.J. Plath; Match 7, (1979) 163.
28. G. Birkhoff, T.C. Bartee; "Angewandte Algebra", R. Oldenbourg Verlag, München, Wien (1973) 264 ff.
29. H. Hermes; "Einführung in die Verbandstheorie", Springer - Verlag, Berlin, Heidelberg, New York (1967) 8.
30. T.L. Gilchrist, R.C. Storr; "Organic Reactions and Orbital Symmetry", Cambridge University Press, Cambridge (1979) 184.
31. E.C. Hass, P.J. Plath; Stud. Phys. Theoret. Chem. 28 (1983), 405.
32. P.J. Plath, E.C. Hass; Stud. Phys. Theoret. Chem. 28, (1983), 392.

33. H. Primas; in R.G. Wooley (Ed.) "Quantum Dynamics of Molecules", Plenum Press, New York (1980) 39.

34. F. Harary; "Graphentheorie", Oldenbourg Verlag, München, Wien (1974) 158 ff.

35. A. Streitwieser; "Molecular Orbital Theory for Organic Chemists", J. Wiley and Sons, New York, London (1961).

36. E. Heilbronner, H. Bock; "Das HMO – Modell und seine Anwendungen", Verlag Chemie, Weinheim (1970).

37. K. Yates; "Hückel Molecular Orbital Symmetry", Academic Press, New York (1978).

38. K. Ruedenberg; J. Chem Phys. 22 (1954) 1878.

39. A. Graovac, I. Gutman, N. Trinajstić; "Topological Approach to the Chemistry of Conjugated Molecules", in "Lecture Notes in Chemistry" Nr. 4, Springer Verlag Berlin (1977).

40. R. Zurmühl; "Matrizen und ihre technischen Anwendungen", Springer Verlag, Berlin (1964) 146 ff.

41. R.B. Woodward, R. Hoffmann; "Die Erhaltung der Orbitalsymmetrie", Verlag Chemie, Weinheim (1970).

42. H.E. Zimmermann; Angew. Chem. 81, (1969) 45; Angew. Chem. Intern. Edit. 8, (1969) 1; Acc. Chem. Res. 4 (1971) 272.

43. E. Heilbronner; Tetrahedron Lett. (1964) 1923.

44. A. Graovac, N. Trinajstić; Croat. Chim Acta 47 (1975) 95; J. Mol. Struct. 30 (1976) 416.

45. M.J.S. Dewar, R.C. Dougherty; "The PMO – Theory of Organic Chemistry", Plenum Press, New York, London (1975).

46. R.B. Woodward, R. Hoffmann; J. Amer. Chem. Soc. 87 (1965) 395, 2046, 2511; Angew. Chem. 81 (1969) 797; Angew. Chem. Intern. Edit. 8 (1969) 781.

47. W. Rinow; "Lehrbuch der Topologie", Deutscher Verlag der Wissenschaften, Berlin (DDR), (1975) 298–302.

48. R. Brdička; "Grundlagen der Physikalischen Chemie", Deutscher Verlag der Wissenschaften, Berlin (DDR) (1965) 527–530.

49. A.M. Butlerov; Z. Chem. 36 (1861) 549.

50. C. Jochum, J. Gasteiger, I. Ugi; Angew. Chem. 92, 7 (1980) 503.

51. H. Michel; "Maß- und Integrationstheorie I", Deutscher Verlag der Wissenschaften, Berlin (DDR) (1978).

52. H. Primas; "Chemistry, Quantum Mechanics and Reduction-

ism", Springer Verlag, Berlin, Heidelberg, New York (1981) 220-244.

53. E.C. Hass, P.J. Plath, Discrete Applied Mathematics 19 (1988) 215-237.
54. W. Schultz-Piszachich; "Tensoralgebra und - analysis", BSB - B.G. Teubner Verlagsges. Leipzig (1977).
55. M.R. Gardner, W.R. Ashby; Nature 228, 21 (1970) 784.
56. K. Göldner; "Mathematische Grundlagen der Regelungstechnik", Leipzig (1968).
57. W. Ebeling, R. Feistel; "Physik der Selbstorganisation und Evolution", Akademie - Verlag, Berlin (DDR) (1982).
58. J.C. Abbott (edit); Trends in Lattice Theory, D. van Nostrand, Reinhold Comp., Canada (1970).
59. G. Kalmbach; "Orthomodular Lattices", London Mathematical Soc. Ser. 18 (1983).
60. E.C. Hass, P.J. Plath; J. Mol. Catal. 29 (1985), 181.

Chapter 4

USE OF OPERATOR NETWORKS TO INTERPRET EVOLUTIONARY INTERRELATIONS BETWEEN CHEMICAL ENTITIES

V.A. Nikanorov and V.I. Sokolov

A.N. Nesmeyanov Institute of Organoelemental Compounds, Russian Academy of Sciences, 117813 Vavilova 28, Moscow, Russia

The operator-network approach (ONA), first proposed by the authors in 1975 to interpret the structural and dynamic interrelations existing between chemical entities for which an evolutionary association can be established, is discussed here. Being essentially a heuristic method, ONA helps one to analyze the internal organization of chemical sets by constructing n-dimensional equifinal and periodic operator-evolutionary networks, based on elementary operators (such as H^+ and e^-) and any number of composite operators. It is shown that many widely recognized chemical regularities arising from different forms of similarity, analogy or correspondence (homology, isoelectronicity, isolobality, etc.) can be consistently interpreted within the framework of ONA. They are viewed as stemming from the universal principle of plurality of isonetwork correspondences extant between chemical structures. ONA is used to formulate concepts for the occurrence of elementary chemical reactions with charge-spin symmetry between the initial, transition and final states being preserved or altered. Comprehensive systematics for virtual substitution reactions are developed. A general heuristic principle is advanced, according to which complete scanning of all the parameters in the corresponding operator networks results in the realization of virtually all possible communications. In conclusion, the usefulness of chemical heuristics is shown to be due to the possibility of superimposing the operator-network formalism on a multidimensional space of experimental data (from thermodynamics, kinetics, etc.) with the aim of ascertaining the selection or prohibition rules for the realization of particular virtual interrelations in concrete chemical systems.

4.1 Introduction

The present work is a further development of the operator approach proposed by us in 1975 for the analysis of genetic interrelations between chemical structures and chemical reactions [1]. It is based on (i) ascertaining the elementary and composite operators responsible for the main types of structural variations in chemical entities (stable molecules, intermediates of chemical reactions and transition states), and (ii) analyzing all kinds of ordered combinations of these operators realized in the form of operator series and operator networks.

A fundamental feature of this approach, as we shall show, is its evolutionary nature. (Chemical evolution, used here in a broadened sense of the word, will be understood as a complete set of all the possible quantitative (metric) changes in chemical entities which lead to a change in their qualitative (to a considerable extent non-metric) properties and features.) In the present context, an operator (designated by a capital letter and the sign $\char94$) is a rule or a mathematical law by which a specified quantity or function constituting an element of a given set is mapped into another quantity or function within some new set. If a certain independent operator $\hat{A}$ is successively applied to a chemical entity, g_0, we arrive at the corresponding evolutionary series (1), i.e.

$$g_0 \xrightarrow{\hat{A}} g_1 \xrightarrow{\hat{A}} \ldots \xrightarrow{\hat{A}} g_n, \text{ where } g_n = n \, \hat{A} \, (g_0) \qquad (1)$$

an ordered population of kindred entities genetically related via the operator law $\hat{A}$. If the operator $\hat{A}$ possesses sufficiently universal properties, then, as a result of its action upon some other chemical entity, h_0, one can obtain a similar evolutionary series: $h_n = -\hat{A} \, (h_0)$. Suppose further, as often occurs in chemistry, that this entity can also be mapped into another independent operator $\hat{B}$ (such that $\hat{B}(g_0) = g_0 \, h_0$). Provided the action of the above two elementary operators $\hat{A}$ and $\hat{B}$ can be successively and additively extended to cover all the elements of the chemical set in question, a linear combination of several elementary evolutionary series is generated and an evolutionary network, as shown in Fig. 4.1, is formed.

With this principle of construction, increasing the number of possible elementary operators to n will clearly lead to the formation

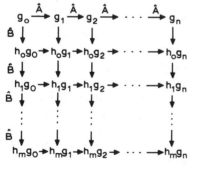

Arbitrary element of network

$$h_i g_j = \left[i\hat{B} + j\hat{A}\right] g_0$$

$$0 \leqslant i, j \leqslant \infty$$

Figure 4.1

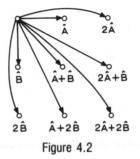

Figure 4.2

of corresponding n-dimensional networks. A characteristic result of all such constructions is the appearance of composite operators as linear combinations of elementary operators. Thus, in the case of the operator network given in Fig. 4.1, these newly generated composite operators, together with the original operators, are shown for the initial portion of the evolutionary network (Fig. 4.2) in the form of a corresponding operator function which is then repeated a certain number of times over the whole network space. Composite operators are actually vectorial sums.

It is thus seen that the operator approach by its very nature makes it possible to consider any chemical entity (network element) in terms of its interrelation with both its nearest and more distant neighbors. We are therefore in a position to describe the present state of such an entity and its previous history, and even to predict its future development, all of which fully conforms to the spirit and methodology of systems analysis [2, 3].

When extended to chemical entities, this approach, conventionally applied in cybernetics [4–6], makes it possible to construct, proceeding from the simpler elements of a chemical set (molecules, particles, reactions) and evolving in accordance with specific operator laws, more complex elements (possibly an infinite number) comprising finite (or n-dimensional) operator-evolutionary networks. Within this framework, the features specifically characteristic of chemical evolution (which distinguish it from evolutionary concepts in other natural sciences) must be determined primarily by the specificity of the selection of the analyzed chemical entities and the concept of their chemical variability. Depending on the selection procedure for the corresponding models, there should exist a certain hierarchy in the possible descriptors of chemical evolution: from the functional (object) level, at which a given entity is described with all its inherent metric parameters, to the increasingly higher levels of generalizing and formalizing its nonmetric (i.e. no longer functional, but operator or matrix) parameters and features.

Accordingly, our object here is to demonstrate the multifarious opportunities afforded by the operator-network approach in the interpretation of chemical substances, involving both the structural (including the steric) and the dynamic evolution of chemical sets. We shall show that numerous characteristic features of the organization of these sets evolve in accord with the laws of the corresponding elementary and composite operators, whose identity and the inter-relations existing between them, can be effectively described in the universal language of operator networks.

4.2 Structural Evolution of Chemical Sets

The need for methods of describing chemical sets that reveal their hierarchic structure and allow one to trace the evolutionary destiny of their component elements is one of the oldest in chemical science. It is hardly accidental that the concepts which in modern language could be justly called operator or network concepts were introduced during the very early stages of the organization of chemical knowledge. Mendeleev's Periodic System of the Elements constitutes an intricately ordered set, each of whose elements could, in turn, be regarded as a kind of supermatrix describing the set of data corresponding to a given chemical element [7, 8]. Network regularities are also clearly revealed in going from elements to the manifold series of their simplest derivatives (oxides, hydrides, etc.) [7]. Similar regularities were established, probably for the first time in the history of chemical science, by Gerhardt. Using organic chemicals as his basis, he constructed the theory of types [9] in the mid-nineteenth century. Subsequent important work involved the investigation of various chemico-algebraic analogies [10–15], and representing substances and reactions in a vectorial space of atomic and molecular components [16–18]. This paved the way for to the present stage of development in this area, which started in the 60s and has been actively pursued up to present [19–34]. In all these studies many possible ways of applying mathematical and formal-logical approaches to the interpretation of chemical – including stereochemical – problems have been considered (often with resort to modern computer techniques).

One of the major current trends appears to lie in establishing a detailed structural organization of chemical sets, since it is the structural peculiarities of chemical particles that determine to a

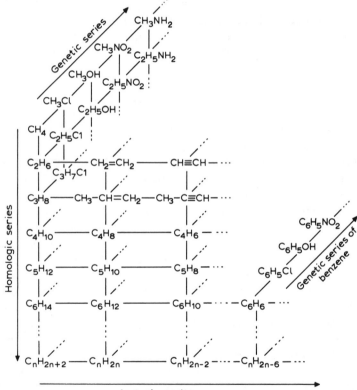

Figure 4.3

considerable extent their reactivity. A clear-cut application of the operator approach to organic chemistry resulted in the principle of homology [35–36], and that has made systematizing a great variety of organic compounds in terms of their homological, isological and genetic series (Fig. 4.3) possible. The homological difference (or homological modulus) $-CH_2-$ is perhaps the first chemical operator; vinylogy, phenylogy, etc. may serve as additional examples. Our approach envisages a further expansion of both the concept of the homological modulus and the homology principle itself.

Such approaches have been employed, for instance, in present-day studies aimed at ascertaining in ever greater detail the hierarchic relations existing between the substituted derivatives of major chemical elements, as undertaken by Hendrickson [22] for the tetracoordinated carbon atom (Fig. 4.4).

Intuitively, it seems evident that this kind of hierarchic inter-relation should be entered into the structure of chemical sets, with not only the nuclei but also the electrons of the particles composing them being taken into account. Let us consider now a major structural theme in coordination chemistry – the evolution of clusters having the general formula EL_n. Each cluster is formed from a central atom of the element E bonded to a number of ligands L (n is the coordination number) and each is characterized by a certain formal

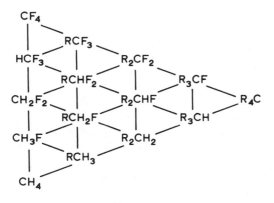

Figure 4.4

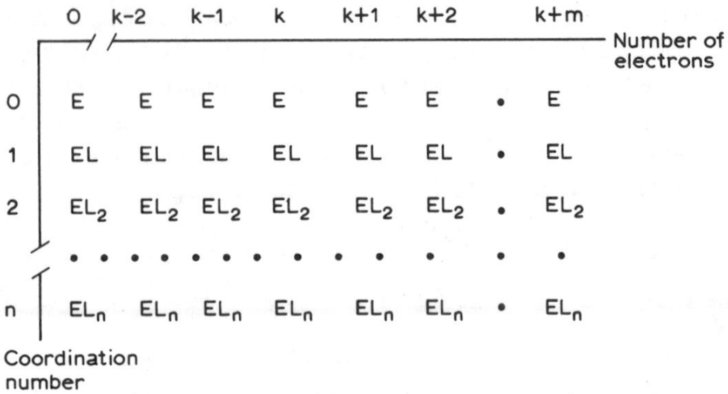

Figure 4.5

electric charge. Since n for every cluster is determined by the number of electrons, k, belonging to the central atom, a group of clusters can be adequately represented as an ordered set on (k, n) coordinates (Fig. 4.5).

This representation, however, does not reveal with suffecient clarity the structural interrelations between the components of our array. That is why we now recast it (having choosen the carbon atom as the element *E* and hydrogen as the simplest ligand) to indicate the formal electric charges on the component clusters, as determined by the valency of the central atom (Fig. 4.6). What attracts our attention then is that the various clusters comprised within our array by no means have the same ability to exist independently, if by this is understood the preservation of their structural individuality.

The central dashed diagonal line denotes neutral clusters, with anions lying above and cations below it. Clusters with a charge that is too high have a small probability of existing under normal conditions. That is why, generally speaking, the only clusters that actually exist are those located in a rather narrow band about the central diagonal line in the array (Fig. 4.6a). Although no sharp boundary can be drawn between these three subsets, there is no doubt that the lower left-hand portion of our array represents the ionic plasma state of matter, where the structural individuality of the clusters is essentially lost. This state is facilitated by an increase

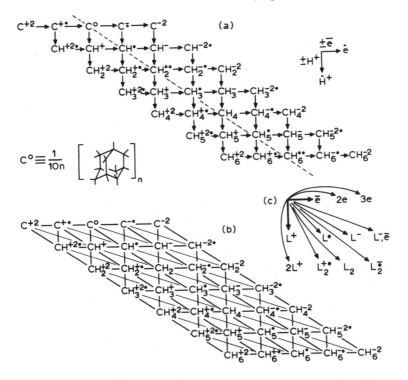

Figure 4.6

in temperature and a decrease in pressure in the system. In turn, the upper right-hand corner of our array represents the metallic state of matter; this is facilitated by a decrease in temperature and an increase in pressure in the system. The structural individuality of the clusters is also lost in this case. Thus, the median subarray of clusters, grouped about the central diagonal line, is of greatest interest in regard to the mutual evolutionary transitions occurring in clusters (Fig. 4.6a).

Initially, we observe that all the clusters in this subarray are strictly and consistently connected with one another by certain operator relations, forming a corresponding operator series, with all the clusters similarly oriented relative to one another and controlled by the same operators.

Only two of the operators shown in Fig. 4.6a are in this case truly elementary and independent (i.e. unmappable into each other). The first is the operator corresponding to one electron $(\hat{e}^-)$ being added to or removed from any cluster, which can be called the redox operator. The other operator involves an acid-base equilibrium $(\hat{H}^+)$ in which one positive ligand (in the simplest case in question – a proton) is added to or removed from the cluster. The evolutionary series obtained from the action of these two elementary operators may be geometrically oriented in a planar network, in the horizontal and the vertical directions, respectively as shown in Fig. 4.6a.

At the same time, as shown in Fig. 4.6b, one can also trace in this array some other evolutionary series oriented along a family of diagonal lines with respect to the above two elementary series. These diagonal series are also realized by the successive action of individual operators (e.g., $\hat{H}^-, \hat{H}, \hat{H}_2, \hat{H}_2^+$, etc.) on the particles in the subarray. The operators, however, are no longer independent but now constitute linear combinations, since they are derivable (Fig. 4.6c) from the elementary operators $\hat{H}^+$ and $\hat{e}^-$. As the procedure for such addition is invariant with respect to both the actual composition of the clusters involved and their arrangement in the network, the resulting combination of elementary and composite operators constitutes a kind of an operator motif (similar to the "wind rose" symbol on medieval geographic maps) applicable to any cluster in the network and is translated all along the network.

The result obtained is clearly of interest from the point of view of its -information theoretical content. At first sight, constructing the evolutionary network shown in Fig. 4.6 from the primary chemical set (Fig. 4.5) seems to contradict a well-known cybernetic theorem which states that the information output from a logical automatic device equals the information input plus the information embodied in the program. Indeed, at the final stage of our logical deduction we did obtain complex chemical processes associated with the transfer of individual electrons and nuclei between the chemical clusters, as well as electrons and nuclei simultaneously, though initially obtaining such additional information was not anticipated. In this connection, however, it should be stressed that our result does not disprove the above theorem but, on the contrary, may be regarded as confirming its general nature for new, purely chemical material. In fact, information on the successive and regular growth in the number of electrons and ligands in the system (Fig. 4.5) (introduced as input for our logical program), together with information regarding the resolution of all kinds of mutual relations and transitions between the clusters in question ("embodied" in the program itself) inevitably ensures exhaustive information as output on the structural evolution of the clusters in the array.

The above formal-logical scheme thus constitutes a chemical algorithm enabling one to describe with the help of only two elementary operators all the possible types of clusters for any given element. This description must be exhaustive as it is based on enumerating all the possible coordination numbers and electric charges for a given element. With this kind of description, not only the particles but also the information on the communications between them appear in the system. The evolutionary networks in question thus make it possible to present simultaneously in a compact form the exhaustively complete equifinal sequences* of the elementary routes of chemical reactions. The most topical examples of current experimental investigation which are also of considerable practical interest are the processes of reductive (Fig. 4.6) and oxidative (Fig. 4.7) "fixation" of carbon** and dinitrogen (Figs. 4.8, 4.9).

---

*The total number of networks possible for each element is generally speaking, infinite, since each component in any given evolutionary network can in principle act either as a substrate or as a reagent (operator) in any other network. The property of the equifinality of these networks is in turn determined by the plurality of the possible ways in which various subsets of clusters and communications between them (evolutionary series) intersect with one another. Hence there arises the possibility of different competitive pathways of generating these clusters and thereby very diverse routes of circuiting the network spaces.

**The initial structural unit with which these schemes start is atomic carbon – the elementary unit of the polymeric allotropic modification of carbon, from which it can be obtained by pyrolysis. Derivatives with coordination numbers 5 and 6 have been known for a long time in the chemistry of the carboranes [37]. In recent years, purely carbonic hexacoordinated clusters have also been discussed [38]. Higher coordination numbers seem to be impossible for carbon.

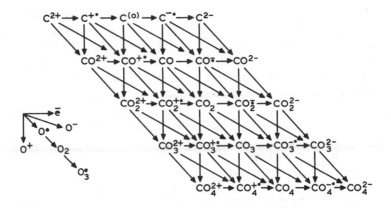

Figure 4.7

When discussing this kind of schematic diagram, one needs to take into account the fundamental duality of the even-odd parity of electons in the central E atom (starting from the central diagonal line). The above operator network constructed for mono-carbon (Fig. 4.6) is representative of the networks for other elements having an equal number of electrons, E_e. The corresponding networks for atoms with an odd number of electrons, E_o, (e.g., nitrogen (Fig. 4.8)) appear at first sight to be completely similar. In the chemical aspect, however, there is quite a substantial difference between them: in the first case the electroneutral particles are $E_2, E_4 ..., E_2$, whereas in the second – $E, E_3, ..., E_{2+1}$. The same relationships will naturally be observed for cations, anions (and all other particles) with a constant difference of one electron for all E_n.

Apart from equifinality, another important property of these operator networks is the periodic character of the changes in the properties and nature of the clusters constituting them. This peculiarity is clearly revealed when such networks are transformed into periodic tables grouped, for instance, in accordance with the principle of similarity of formal electric charge for corresponding clusters, as shown in Fig. 4.10 for even (represented by carbon) and for odd (represented by phosphorus) elements. Such tables of

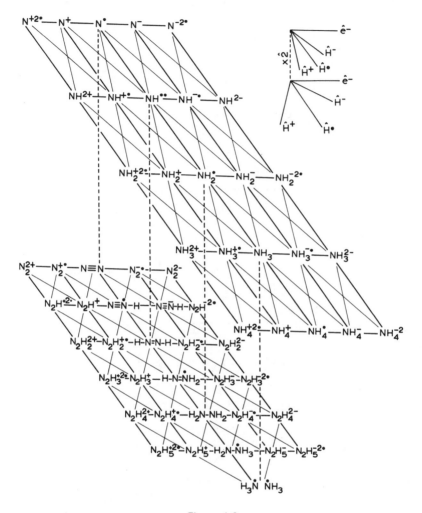

Figure 4.8

intermediates can be regarded as a kind of periodic system for the corresponding chemical reactions.

The number of even the simplest chemical operators is very large (possibly infinite). The "oxygen" operator (Figs. 4.7, 4.9), the

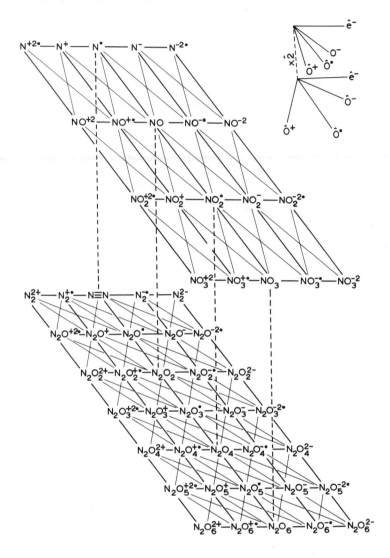

Figure 4.9

C^{+2}	$P^{+2\bullet}$	$C^{+\bullet}$	P^+	$C^{\bullet\bullet}$	$P^\bullet$	$C^{-\bullet}$	P^-	C^{-2}	$P^{-2\bullet}$
PH^{+2}	$CH^{+2\bullet}$	$PH^{+\bullet}$	CH^+	$PH^{\bullet\bullet}$	$CH^{\overline{\bullet}}$	$PH^{-\bullet}$	CH^-	PH^{-2}	$CH^{-2\bullet}$
CH_2^{+2}	$PH_2^{+2\bullet}$	$CH_2^{+\bullet}$	PH_2^+	$CH_2^{\bullet\bullet}$	$PH_2^\bullet$	$CH_2^{-\bullet}$	PH_2^-	CH_2^{-2}	$PH_2^{-2\bullet}$
PH_3^{+2}	$CH_3^{+2\bullet}$	$PH_3^{+\bullet}$	CH_3^+	$PH_3^{\bullet\bullet}$	$CH_3^\bullet$	$PH_3^{-\bullet}$	CH_3^-	PH_3^{-2}	$CH_3^{-2\bullet}$
CH_4^{+2}	$PH_4^{+2\bullet}$	$CH_4^{+\bullet}$	PH_4^+	CH_4	$PH_4^\bullet$	$CH_4^{-\bullet}$	PH_4^-	CH_4^{-2}	$PH_4^{-2\bullet}$
PH_5^{+2}	$CH_5^{+2\bullet}$	$PH_5^{+\bullet}$	CH_5^+	$PH_5^\bullet$	$CH_5^\bullet$	$PH_5^{-\bullet}$	CH_5^-	PH_5^{-2}	$CH_5^{-2\bullet}$
CH_6^{+2}	$PH_6^{+2\bullet}$	$CH_6^{+\bullet}$	PH_6^+	$CH_6^{\bullet\bullet}$	$PH_6^\bullet$	$CH_6^{-\bullet}$	PH_6^-	CH_6^{-2}	$PH_6^{-2\bullet}$

Figure 4.10

operator of dimerization (Figs. 4.8, 4.9), etc., serve here as relevant examples. In describing more complex relations, it may be necessary in the general case to introduce new operators and to change over from planar to n-dimensional networks. An example of this is illustrated in the three-dimensional networks shown in Figs. 4.8, 4.9 for the redox reaction of nitrogen (third operator: association-dissociation). Another example is provided by the three-dimensional network for organometalic compounds shown in Fig. 4.11, where the difference between π- or σ-bonding of ligands L with the metal atom is accounted for. It is of interest to note that specific cases of such networks are already frequently used in experimental studies to

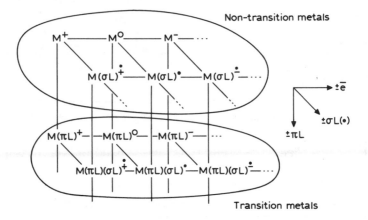

Figure 4.11

demonstrate the reactivity of organic and organometallic compounds (one of the numerous examples, taken from [39], is shown in Fig. 4.12).

The possibility of different types (π or σ) of bonding between the ligands and the central atom in organometallic compounds raises the problem of accounting for dissimilar populations of the metal atom electron orbitals in the evolutionary networks in horizontal series of clusters corresponding to the operator L. In describing the various valency and coordination states in Fig. 4.12 these differences are not shown. In principle, however, they should always be borne in mind, if the network approach is to model not only the charge and coordination peculiarities of the clusters in question but also the differing nature of their chemical bonds and electron shell populations.

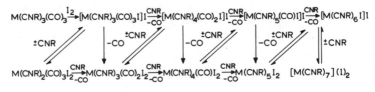

Figure 4.12

Strictly speaking, every cluster (represented as a point in the evolutionary network) has a corresponding set of bonding, non-bonding and antibonding orbitals, i.e. this point can be regarded as a matrix in a multidimensional space. However, to a fair approximation, describing the structural interrelations between the clusters can in general be accomplished without taking into account all these orbitals, and it is often sufficient to indicate only those which, in energetic as well as in steric and symmetry respects, play the most important role in the chemical reactions of the clusters and determine their structural individuality. For instance, to understand the evolutionary interrelations and the nature of reactivity of the clusters in the evolutionary series of monocarbon (operator H^+), namely:

$$CH_4 \xrightarrow{-H^+} CH_3^- \xrightarrow{-H^+} CH_2^{2-} \xrightarrow{-H^+} CH^{3-} \xrightarrow{-H^+} C^{4-}$$

increasing electron-donor nature

it suffices to consider from the set of the orbitals present (shown in Fig. 4.13 [40], only those higher populated boundary [41] orbitals, for which the dashed line shows a tendency to an increase in the energy level.

The language of operator-evolutionary networks thus proves itself extremely suitable for expressing with maximum brevity

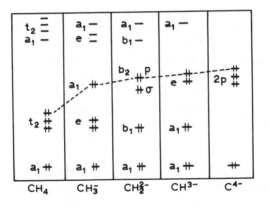

Figure 4.13

quite voluminous and substantial chemical information schemata. Moreover, it is well suited for ascertaining the basic principles of the structural organization of chemical sets and the genetic interrelations existing between practically all chemical entities (molecules and particles), with all the various communications between them being accounted for.

We emphasize here that in many of the studies devoted to ascertaining a certain degree of identity between diverse chemical entities authors often use in implicit form a variant of an operator-network description of chemical reality. Indeed, any chemical comparison can be reduced to one of two main types: the mapping of entities belonging to the same network set (which can be called an intranetwork comparison) or the mapping of entities belonging to different network sets (an internetwork comparison)*. In the first case the mapping actually establishes the laws governing the composition of a given network, i.e. the elementary and composite operators actualizing the communications between the elements of the set. The final result of mapping in the second case will be determination of the common features of several examined networks, which in turn makes it possible to represent them as subsets within a more comprehensive chemical set.

One could cite numerous examples (taken from different spheres of chemistry) of using "networks", and of comparisons and mappings of the above two types. A typical example of "intranetwork" comparison is the use of the already mentioned principle of homology [35, 36], starting with variants that have long been known – vinylogy, phenylogy, etc. (Fig. 4.14). In its most general form, the principle of isomorphic substitution [16b] allows for the insertion of very diverse homological moduli into the vertices, edges and faces of plane and polyhedral (Fig. 4.15) structures [42]. It also seems possible to regard the construction of "trefoil" aromatic systems [43], and the transition states of chemical reactions

---

*What we call a "network" comparison (as distinct from the conventional non-network or "functional" one) is a comparison where the entities are mapped not in their isolated form but rather as the elements of the evolutionary series corresponding to them.

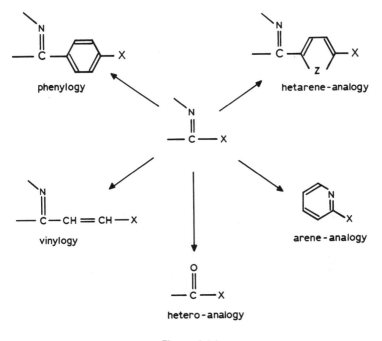

phenylogy

hetarene-analogy

vinylogy

arene-analogy

hetero-analogy

Figure 4.14

catalyzed by metals (metalalogy)[44] as variants of homology.

Mention may also be made of well-known examples of various kinds of mappings of elements from different chemical sets, which can be compared with each other because they possess certain common features. Two trends can be traced here now – to establish these interrelations at the electronic and structural levels, with progress being determined in both cases by the selection of new criteria for the mapping of chemical entities. For the electronic level we note use of the isoelectronicity principle [45], establishing the correspondence between the clusters (of different elements) in terms of their total number of electrons, the electron equivalence principle [46], establishing this correspondence by the number and nature of valency electrons, and, finally, the use of the isolobal analogy principle [40, 47], founded on the similarity of the number, symmetry,

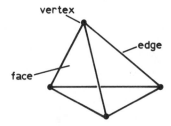

isomorphous substitution

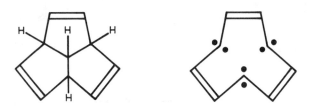

trefoil aromatic systems

Figure 4.15

properties, shape and energy of boundary orbitals in element-centered clusters (Fig. 4.16).

Also worthy of attention is a variant of combining the electronic and structural levels when comparing chemical entities, manifested most clearly in the isoconjugation principle proposed by Dewar [48]. This establishes analogies between molecules having an equal number of atoms in the conjugated chain, the same type of bonding, and the same number of π-electrons (which presupposes their having molecular orbitals of the same size and shape, formed from the same number of individual atomic orbitals) (Fig. 4.17).

It is of interest to note that, in the terminology employed here, the principle of isolobal correspondence can be regarded as establishing the isomorphism between individual elements of different evolutionary networks (i.e. between individual derivatives of isoelectronicity: H, He$^+$, Li^{2+}, Be^{3+}

electron equivalency:

17e: Co(CO)$_4$ ~ Mn(CO)$_5$• ~ CpFe(CO)$_2$• ~ I•
16e: Fe(CO)$_4$ ~ Cr(CO)$_5$ ~ S
15e: Co(CO)$_3$ ~ Mn(CO)$_4$ ~ V(CO)$_5$ ~ CpFe(CO) ~ N

isolobality:

CH^3 ~ M(CO)$_5$ M = Mn, Re
CH$_2$ ~ FeL$_4$ ~ PtL$_2$
CH ~ CpM(CO)$_2$ ~ Co(CO)$_3$ M = Cr, Mo, W

Figure 4.16

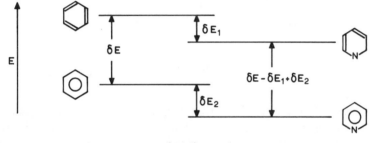

Iso-conjugation

Figure 4.17

different chemical elements). Generally speaking, this does not automatically presuppose a similar isomorphism between the corresponding evolutionary series or evolutionary networks as a whole. From the fundamental duality, namely the even-odd parity of the electrons of E atoms in EL_n clusters (noted above when comparing Fig. 4.6 and Fig. 4.8), it follows that basically only two types of isolobal correspondence can exist in the Periodic System of the Elements: those alternating and nonintersecting series (Fig. 4.18) of EL_n clusters having respectively an even and odd overall number

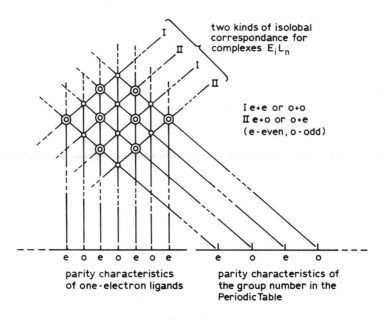

two kinds of isolobal
correspondance for
complexes $E_i L_n$

I e+e or o+o
II e+o or o+e
(e-even, o-odd)

parity characteristics
of one-electron ligands

parity characteristics of
the group number in the
Periodic Table

Figure 4.18

of electrons. Moreover, from this it may be concluded that isolobal correspondences between EL_n clusters with an even overall number of electrons and those with an odd overall number of electrons are in principle prohibited.

Apart from the above structural and electronic factors, on which it is now customary to perform all kinds of comparisons between chemical entities (Figs. 3, [14–17]), it often proves necessary or useful to map chemical entities and the interrelations existing between them by considering other possible categories. In the present context, this "plurality" is by no means accidental but constitutes a natural consequence of the fact that the same chemical entities may be simultaneously regarded as the elements of a great variety of operator-evolutionary networks, depending on the nature of the elementary operators used in a given situation. In this case, the laws governing the constructing of any kind of operator-evolutionary chemical network (based on the elementary operators examined by us or on other elementary operators) and the informational content contained within them should be of a universal nature. Then, all the numerous "iso-principles" of chemical affinity, similarity or equivalence so far proposed can be regarded as individual and important corollaries of a more general and unified principle. Accordingly, for the manifold possible operator-evolutionary networks constructed from chemical clusters, proceeding from particular combinations of different elementary operators, there will always exist a set of general features and categories. From this set one can establish one or other of the known (one-to-one, one-to-many, many-to-one, many-to-many) types of correspondence for the clusters forming these networks. This principle will be called the principle of the plurality of known correspondences.

An independent and very important sphere for the application of operator-network approaches is represented by evolutionary stereochemistry, where the structural-electronic level of mapping chemical properties and their mutual transmutations are considered in terms of their spatial evolution, i.e. the variation of their symmetry and configurational characteristics (the latter for chiral structures) is considered.

In the operator-network form, one can present, for instance, symmetric interrelations arising from ligand substitution in coordination polyhedra, shown in Fig. 4.19 for the case of a tetragonal pyramid. In this case, stereochemical evolution is determined by the reduction of the symmetry levels of the clusters in question, and is shown by mapping the corresponding point groups onto the apices of the operator network. The operator for transition from one point to another in a given network is the substitution of ligand A by ligand B. The effect of its action, however, depends on the position occupied by the substituted ligand: substitution of A in the basal plane of an A_5 molecule yields the A_4B^* structure of C_S symmetry, whereas a substitution of the ligand in the apical position results in the A_4B molecule of C_{4v} symmetry (see Fig. 4.19). This indicates that the network in question is not described un-ambiguously by the $A \rightarrow B$ substitution operator, and in this sense is not linear. Using this principle, as demonstrated further in Fig. 4.20 for five other simple polyhedra, structural-symmetric interrelations can also be represented for any other

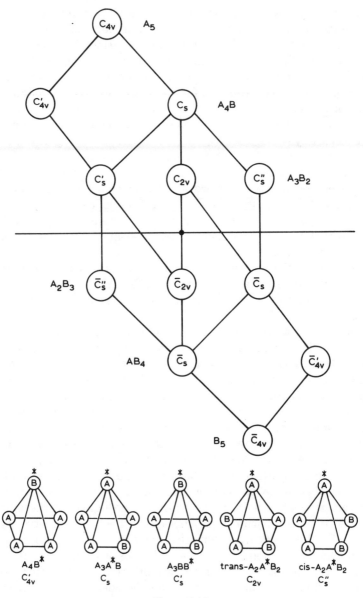

Figure 4.19

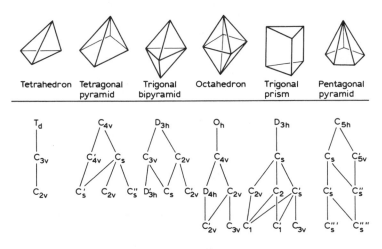

Figure 4.20

polyhedral system. Schemes of this kind can be regarded as special types of stereochemical graphs adequately characterizing the corresponding polyhedra in their evolutionary series. This makes the schemes useful for the analysis of stereochemical problems associated with studying the properties of heteroligand complexes, the symmetry properties of reaction routes*, etc.

For the operator-network treatment of configurational stereo-chemical relations it is in turn essential to draw a clear-cut boundary between achiral and chiral structures. These relations, exemplified for tetracoordinated tetrahedral derivatives by means of successive substitution of ligands, are shown in Fig. 4.21. This figure depicts the pathways of successive transitions from an elementary achiral cluster to an elementary chiral cluster *EABCD* brought about by substitution with achiral ligands *A*, *B*, *C*, and *D*. Below the dashed

---

*In the case of chiral point groups appearing in the network the routes will naturally bifurcate into two enantiomeric channels (for the structure of a trigonal prism of C_2 symmetry in Fig. 4.20 only one of them is shown).

(a)

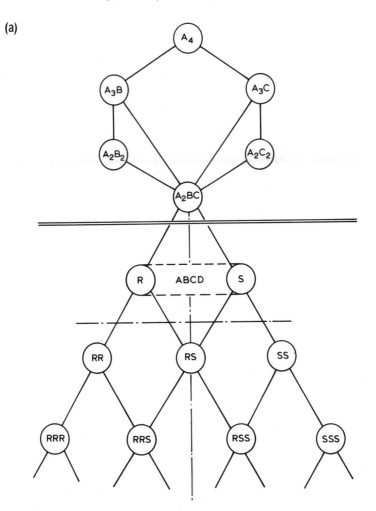

Figure 4.21(a)

line in Fig. 4.21, we also show the further elaboration of the initially chiral molecule, resulting from the introduction of a new operator providing for progressive growth in the number of chiral centers. Depending on the number of structurally identical elements of the

(b)

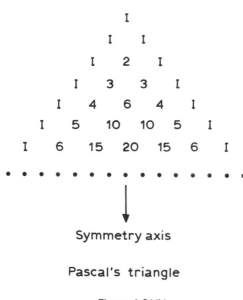

Symmetry axis

Pascal's triangle

Figure 4.21(b)

same or opposite absolute configuration (R or S), the molecule will exist in the form of optically active or optically inactive (meso)-diastereomers. Such notation, using all possible sets of configuration descriptors as the vertices of the graph (the differences in the mutual arrangement of the chiral centers in Fig. 4.21a are not presented in detail), produces an operator-network expression for all the main stereochemical notions characterizing the differences between stereoisomers. Thus, clusters symmetric with respect to the central line are *enantiomerically* related, those arranged in one row are *diastereomerically* related, and those next to each other in one row are *epimerically* related. We note here the relation of this network to probably the oldest evolutionary network – the Pascal triangle (Fig. 4.21b). Both the similarities and the differences between them are of interest. Pascal's triangle has a single initial point: the vertex and a central symmetry axis. The network of chiral molecules also has a symmetry axis (on which lie for even-numbered series all the mesoforms having the same number of chiral centres of the opposite configuration), but the single vertex (and other similar points located at the intersections of the odd-numbered series with the central axis) now becomes two because of the appearance of corresponding enantiomeric pairs.

The immediate vicinity of the origin of coordinates is comprised of 12 prochiral molecules of A_2BC type (in Fig. 4.21a for simplicity only one of them is shown). The complete graph for successively substituted vertices in the tetrahedron, including $ABCD$, has 35 vertices. In its rows the chiral network reflects the structure of the Pascal triangle (Fig. 4.21b) by the manner in which the elements are combined, though there are no numerical (binomial) coefficients. Along its diagonal lines two constant operators, R and S, act, while along the diagonal lines (except the two extreme ones) in the arithmetical triangle the operators do not remain constant but evolve according to a linear law.

The material presented in this section shows that operator-network concepts find wide and multifarious application in various spheres of chemistry. These concepts represent a sufficiently universal language to make it possible to interpret a number of problems involved in investigating different kinds of analogies and correspondences between chemical entities when their electronic and structural peculiarities (including regio- and stereoisomerism) are taken into account. The following section illustrates the use of these concepts for the analysis of some chemical dynamics problems related to the reactivity of molecular systems.

4.3 Evolution of Reaction Mechanisms

In the previous section the evolution of chemical sets was considered mainly from the viewpoint of systematic changes in the chemical structure and composition of the entities forming the set. However, as shown above, the operators that are responsible for such chemical changes conform to the real chemical reactions associated with the transfer of electrons and/or nuclei onto the reacting substrate. The operators also determine the type of reaction taking place, its molecularity, as well as the nature of the transition states and the intermediates formed in the course of the reaction.

Since all these concepts form part of physical organic chemistry, whose primary objective lies in investigating reaction mechanisms, we now consider separately the question of the evolution of the types and mechanisms of reactions in a chemical set represented by a complete operator network.

At the present time, physico-organic chemistry favors a definite systematics for substitution reactions which appeared and acquired its current form as a result of a large number of experimental and theoretical studies initiated by the British chemical school (Ingold, Hughes, *et al.*). The systematics is based on subdividing reagents into electrophilic (symbol E), nucleophilic (symbol N) and homolytic (or radicophilic) (symbol H), and the substitution reaction themselves into monomolecular (symbol 1) and bimolecular (symbol 2) [49, 50]. According to this classification, a molecule of (say) a typical organic substrate $C - X$ (where X is any organic or organoelemental function: H, Hal, Alk, Ac, OR, metal, etc.) must be associated with nine principal processes, *viz.* three monomolecular and three pairs of bimolecular reactions of substitution at atoms C or X.

Another variant in classifying the mechanisms of these reactions [51, 52], is based on the nature of the attack of the substrate (symbol A, $\mathscr{F}$) and the removal of the leaving group (symbol D, $\mathscr{F}$) in the intermediate or in the transition state of the process. It should be noted that the classical systematics of mechanisms are based on an unstated postulate implying a coincidence of the natures of the elementary actions A and D during the corresponding substitution processes. According to this classification and the above postulate*, in heterolytic substitution reactions only ionic (but not radical) particles are generated, and in homolytic substitution reactions only free radicals (by no means ions) are formed. Since the entities corresponding to the initial and final states of reactions have in this case identical formal electronic charges (or spin multiplicity), all the processes considered can be regarded as taking place

---

*This postulate can be regarded as a corollary of the principle developed by Lewis and London, according to which the course of concordant thermal processes is adiabatic and not accompanied by electron excitation [49].

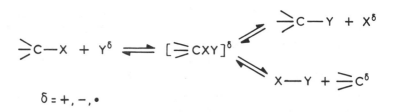

Figure 4.22

with conservation of charge-spin symmetry (with respect to the transition state or the intermediate), as shown in generalized form in Fig. 4.22.

The above systematics of substitution reaction mechanisms are automatically deducible from the complete chemical set of entities for any atom (constructed by us above). To demonstrate this, it is sufficient to note that the universal operator for all the ligand-electron networks described in the preceding section precisely corresponds to all the classical mechanisms for mono- and bimolecular substitution at a saturated atom. To facilitate understanding of this operator, and the canonical designations of the corresponding mechanisms, we illustrate it in Fig. 4.23 for a typical stable hydrocarbon (the methane molecule). Naturally, the conclusions arrived at can be readily extended to include any other $E-X$ bonds and A, D reagents.

However, further mapping of the links showing all possible (so-called virtual) pathways for the mutual interconversion of clusters within a comprehensive operator network suggests that the classical systematics presented in Fig. 4.23 by no means cover all the mechanisms of elementary substitution reactions. Indeed, what we have called the "classical" operator of substitution belongs not only to the methane molecule but also to any other cluster in the network, specifically to the CH_4^+ cationic radical (Fig. 4.24a) and the CH_4^- anionic radical (Fig. 4.24b). In this case, however, upon mapping Figs. 4.23 and 24, it is evident that one can observe as it were an intranetwork of "mutual intersection" of the classical substitution mechanisms corresponding to all three entities (CH_4, CH_4^+ and CH_4^-). This then results in the formation of a series of

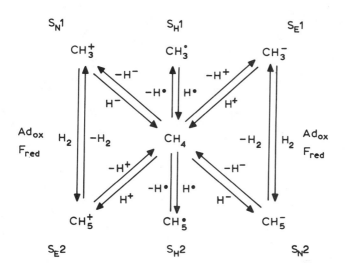

Figure 4.23

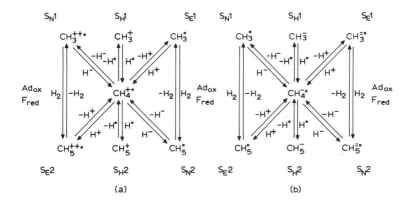

(a)   (b)

Figure 4.24

"non classical" variants of substitution where the natures of the elementary actions of attack (A) and removal (D) do not necessarily have to coincide.

Realization of this possibility is shown in Fig. 4.25. A characteristic feature of all the non-classical mechanisms considered here

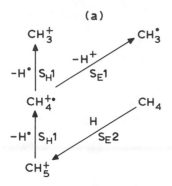

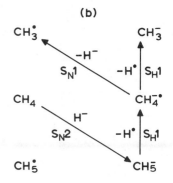

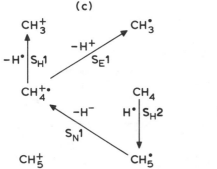

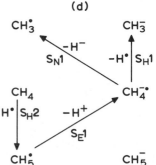

Figure 4.25

is the change in charge-spin symmetry between the initial and final states during substitution, i.e. the generation of radical particles in ionic reactions (Fig. 4.25a, b) or ionic particles in radical transformations (Fig. 4.25c, d).

Our conclusion can be extended to include all other substrates for the $E-X$ bond and Y reagents. We emphasize that all the mechanisms are bimolecular, and presuppose the transfer of the corresponding ligands (X or E) from the substrate to the reagent Y together with its electrons. Such generation of radicals in ionic processes (Fig. 4.25a, b), according to Pryor's classification [53], should be assigned to so-called "molecularly induced homolysis". By analogy, the formation of ions in radical reactions (Fig. 4.25c, d) may be termed "molecularly induced heterolysis". These two reaction types are currently very important in both experimental and theoretical studies. It is sufficient to mention in processes of radical and ionic cluster formation in organometallic heterolytic demetalation reactions [54, 55], related investigations of the mechanism known in the literature as monomolecular radical-nucleophilic substitution and designated by the symbol $S_{NR}1$ [56, 57], and development of the idea of ionic contributions to radical transition states [58]. Despite the extremely great variety of all the possible processes of this kind in terms of both the initial substrates and the reagents, the operator-network approach leads to (i) complete formal systematics for each, thereby allowing one to make heuristic predictions of new "anomalous" reactions of this type and (ii) a unified genetic interpretation based on their common feature – a change in charge-spin symmetry during the reaction.

Within the framework of a complete operator network, the above mechanisms of nonclassical substitution constitute a fully fledged alternative to the redox mechanism involving individual electron transfer (the SET mechanism [59]). The latter is determined by the action of the $\hat{e}^-$ operator upon the substrate and, as shown in Fig. 4.26, must result in the formation of the same "anomalous" products. The main difference between nonclassical substitution (Fig. 4.25) and the SET mechanism (Fig. 4.26) seems to be a differing degree of separateness of the substrate and the reagent molecules in the elementary act of reaction. Nonclassical substitution (as well as its classical prototypes (Fig. 4.22)) is a process of "tight" donor-acceptor contact of the outer orbitals of reagents ($\overset{|}{\underset{|}{>}}C-X$ and

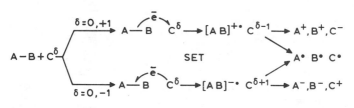

Figure 4.26

$E-D$), ultimately resulting in the formation of a wholly unified molecular orbital of the transition state or the intermediate, where the "individuality" of the electons populating it is essentially lost. The SET mechanism must in turn presuppose a sufficiently "loose" contact of these orbitals, whenever they are not wholly unified, and "individualized" electron transfer is actualized over a sufficiently long distance through space or via the solvent [60]. According to this criterion, the probability of the SET mechanism will be greatly increased in reactions involving participation of sterically hindered systems whose steric features interfere with the "tight" contact of reacting molecules in the transition state of the process.

The coexistence of two types of substitution reactions (proceeding with or without preservation of charge-spin symmetry) within a unified heuristic scheme does not imply an equal probability of their taking place in the corresponding real reaction systems. Thus, for most chemical reactions, proceeding under normal thermal conditions and leading to the formation of an activated complex or intermediate in the electronic ground state, the requirement of charge symmetry preservation seems to be very significant and to determine the composition of the products. In accordance with conventionally accepted terminology, these reactions are concordant and require a definite mutual geometric orientation of the interacting molecules, determined by the principle of orbital symmetry preservation [61]. However, if for any reason (e.g. the presence of steric hindrance preventing a favorable attack or the proximity of the ground and the excited states of a reagent) the process involves the formation of an electron-excited transition state (non-adiabatic mechanism), the requirement of charge-spin symmetry preservation no longer applies, and the composition of the products can then correspond to the typical combinations presented in Figs. 4.25 and 4.26.

The situation examined above for bimolecular substitution reactions, associated with the formation of differing products depending on the conditions under which the reaction proceeds (adiabatic or non-adiabatic), seems to be of a sufficiently general nature that it is observed for many other types of chemical conversions. It is important to emphasize here that all such situations can be very conveniently considered with the help of special operator-evolutionary networks that take into account not only the ground but also the excited states of reacting molecules. To illustrate the construction of such networks, Fig. 4.27 shows in a three-dimensional operator space the well-known alternation of the stereochemical results of electrocyclization and cyclodecomposition reactions, depending on the number of olefin electrons and the conditions under which the reactions proceed (thermal or photochemical), based on Woodward–Hoffmann rules.

Apart from geometrical differences, an important role in the substitution process going via either route can also be played by factors such as the nature of the substrate and reagent, the solvent, the presence of a catalyst, physical effects acting upon the reacting system (temperature, light, etc.). Comparison of the formal operator schemes advocated here with experimental chemical reality thus shows that all of them should be regarded only as maximal generalizations of possible real chemical situations. Corresponding to every entity (represented above as a fixed point in an evolutionary network*) there is thus a multidimensional space describing all the possible geometrical and energetic parameters. The general heuristic principle is based on an exhaustive variation of these parameters for all the possible structures of a given nuclear-electron composition, and realization of the experimental conditions in which the corresponding operators act on real reaction systems. Chemical reality is adequately reflected in these schemes inasmuch as they are based on an exhaustive enumeration of the coordination numbers and available electrons possible for this kind of entity. For the substitution reactions considered in this section, for instance, with the above scanning and varying of the experimental conditions and the parameters of the reacting molecules $\overset{\rightharpoonup}{\nearrow}C{-}X + Y^k (k = 0, {}^{\pm}1)$, all the possible communications between the entities in the network can be realized.

Presently available chemical experience in many cases confirms

---

*This formal-logical equivalent of activated complexes or the intermediates of chemical reactions has been designated as a "transition" in [60]. The procedure of varying the topology of four-unit transitons is described in [62].

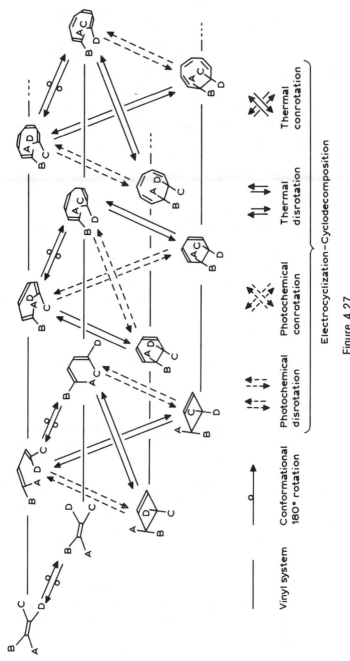

Figure 4.27

this principle, which can thus serve as the starting point for the heuristic prediction of new chemical reactions and the interpretation of their mechanisms.

4.4 Conclusions

The above examples show that operator-network approaches are often useful not only for the interpretation of chemical material but also provide possibly the best representation of vast and substantive arrays of chemical information in compact form with explicit formulation of the laws governing the constructing of diverse chemical sets.

Ascertaining for a given chemical system all the formally possible (virtual) interrelations between its elements, or showing the regularities of their organization within the operator-network approach, does not automatically tell us which of them will actually take place in a real (and not a logically formalized) chemical system and what effect on these interrelations the specific nature of the entities will have. Such internal, or inherent, restrictions on formal operator-network methods are by no means accidental, but quite natural and typical of all heuristic approaches to solving chemical problems. This regularity constitutes a general principle because no formal-logical system by itself (i.e. using only its own means) can solve the problem of the comprehensiveness and truth of its own conclusions. This inference, in our opinion, follows from Goedel's theorem [63] on the incompleteness of any axiomatic theory, which is a well known result in mathematical logic.

There is only one way to circumvent Goedel's restriction as applied to formal-logical approaches in chemistry: to resort to *non formal* physical (or mathematical*) rules, generalizations and regularities (independently derived from either purely theoretical or practical experience). Such rules play in this case the role of

---

*The above-mentioned Woodward–Hoffmann rules (Fig. 4.27) can serve as a vivid example of using a fundamental mathematical category of symmetry to derive chemical rules of selection.

metasystems with respect to the formal-logical approaches. Only they make it possible to select from the full array of logically (virtually) possible processes those capable of being actualized under given real conditions. Bearing in mind the importance of this issue for the correct understanding of the methodological role of operator-network concepts in the general framework of methods used for the theoretical interpretation of chemical material, the most valuable present trend seems to be to use these concepts in combination with physical methods of analyzing and generalizing chemical data, in which numerical or functional parameters are mapped into chemical operators.

In examining the problem from this point of view, the applicability of universal network approaches to practically all spheres of modern chemistry is demonstrated. For instance, the classical Born–Haber cycle in thermodynamics is no more than a fragment of an operator-network furnished with energy characteristics for each transition. Network concepts are also widely used in chemical kinetics to derive with the aid of graphs complex kinetic equations for multistage chemical reactions, where the individual routes in the networks are characterized by their rates [62]. In electrochemistry, the construction of operator networks makes it possible to follow the variations of redox potentials in conjunction with chemical processes (Fig. 4.28) [64]. Rudenko's concept of evolutionary catalysis, formulating thermodynamic, kinetic, probabilistic, and informational criteria for the evolutionary development of open, metal-containing catalytic systems [65], is also associated with a network approach.

Highly fruitful opportunities for combining heuristic operator-network approaches with substantive physical methods are available in various fields of modern stereochemistry. Here conformational calculations make it possible to select the actually realized routes in the internal rotation of molecules or the reorganization of their coordination polyhedra from a complete set of their possible pathways. Study of the kinetics and thermodynamics of particular reactions, in which certain conformers can take part, allows an estimation of their relative reactivity in the evolutionary series of compounds, as shown in Fig. 4.29 for two conformational series of electrically reduced members of the bianthrylidene series. An

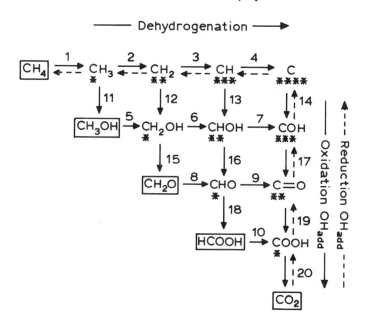

Figure 4.28

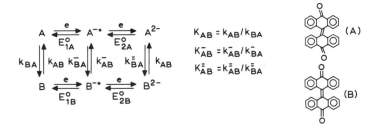

Figure 4.29

evolutionary topological approach of fundamental importance in understanding the significance of steric effects in dynamic stereo-chemistry has recently been demonstrated by Dubois, Panaye et al. [32]. They found three differing regions in the hyperstructure expressing the evolution of the steric parameter E of alkyl groups geomorphic to t-Bu_3C- as a function of their substitution (Fig. 4.30);

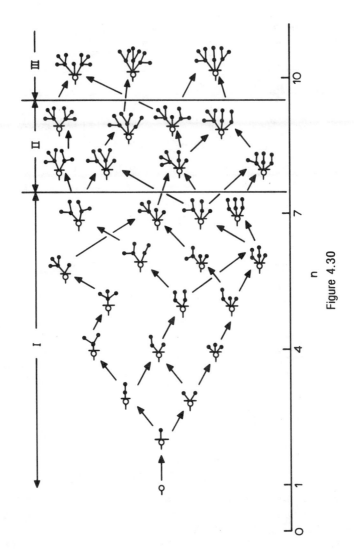

Figure 4.30

this enables one to substantiate the prediction of other possible steric effects.

An important current trend in the development of evolutionary network approaches consists in using them to understand various biochemical transformations, as exemplified by the thermodynamic analysis of major metabolic cycles, and the construction of maps and tables of biochemical metabolic processes, systematized in accordance with the periodicity principle [66].

Although the nature of these transformations is sufficiently complicated, many of them are based on essentially the same fundamental operators as those of organic chemistry ($\hat{e}, \hat{H}^+, \hat{O}$, etc.) used in substitution, addition, elimination, and redox reactions. The selection of particular regio- and stereospecific routes is actualized by the action of high molecular mass catalysts – enzymes, whose structure is in turn a product of prolonged biochemical evolution.

In summary, further expansion of the use of operator-evolutionary approaches in different spheres of chemical science is to be expected. The most promising will likely be the ascertaining of new network interelations between chemical entities with the aim of finding new selection and prohibition rules in series of virtually possible processes and phenomena. Since the operator-network concepts we have considered constitute an integral part of formal-logical approaches to solving chemical problems, the above-mentioned trend of superimposing network chemical operator formalisms on a multi-dimensional spaces of physical parameters will certainly further stimulate the development of chemical heuristics.

4.5 References

1. V.I. Sokolov and V.A. Nikanorov, Zh. Str. Khim., No. 6, 1068 (1975).
2. Bertalanffy L. von, General System Theory, New York, 1968.
3. V.N. Sadovsky, Osnovaniye obshchey teorii sistem (Fundamentals of the General Theory of Systems), Moscow, Nauka, 1974.

4. N. Wiener, Cybernetics, transl. from Eng., Moscow, Sovetskoye radio, 1958.

5. Neumann J. von, The Theory of Self-reproducing Automatic Devices, transl. from Eng., Moscow, Mir Publ., 1971.

6. N.A. Krinitsky, Algoritmy i roboty (Algorithms and Robots), "Radio i svyaz'" publ., 1983.

7. S.A. Shukarev, Neorganicheskaya khimiya (Inorganic Chemistry), Moscow, "Vysshaya shkola" publ., v. 1, 1970, v. 2, 1974.

8. D.N. Trifonov, O kolichestvennoy interpretatsii periodichnosti (On the Quantitative Interpretation of Periodicity), Moscow.

9. C.F. Gerhardt, Traité de chimie organique, t. 1-4, Paris, 1854-1860.

10. A. Cayley, Phil. Mag., 47, 444 (1874); Ber., 6, 614 (1875).

11. J.J. Sylvester, Amer. J. Math., 1, 64 (1878).

12. W. Clifford, Amer. J. Math., 1, 64 (1878).

13. I.O. Dmitriyev and S.G. Semenov, Kvantovaya khimiya - yeyo proshloye i nastoyashcheye (Quantum Chemistry - its Past and Present), Moscow, Atomizdat, 1980.

14. V.G. Alekseev, Zh. Russk. Fiz.-Khim. Obshch, 33, sect. 1, 314 (1901).

15. D.H. Rouvray, Chem. Brit., 10, 11 (1974); Chem. Soc. Rev., 3, 355 (1974).

16. A.A. Balandin, Multipletnaya teoriya kataliza (The Multiplet Theory of Catalysis), pt. 3, MGU Publ., 1970.

17. R. Aris, Ind. Eng. Chem., 61, 17 (1963).

18. P.H. Sellers, Proc. Natl. Acad. Sci., 55, 693 (1966); SIAM J. Appl. Math., 15, 13 (1967).

19. A.T. Balaban (Ed.), Chemical Applications of Graph Theory, London, Academic Press, 1976.

20. A. Panaye, J.A. MacPhee and J.E. Dubois, Tetrahedron, 36, 759 (1980).

21. A.T. Balaban, D. Farcasiu and R. Banica, Rev. Roum. Chim., 11, 125 (1966).

22. J.B. Hendrickson, J. Amer. Chem. Soc., 89, 7047 (1967); 97, 5763, 5784 (1975).

23. E.J. Corey and W.T. Wipke, Science, 166, 178 (1969).

24. W.T. Wipke and T.M. Dyott, J. Amer. Chem. Soc., 96, 4825, 4834 (1974).

25. J. Dugundji and I. Ugi, Top. Curr. Chem., 39, 19 (1973).
26. I. Ugi, J. Bauer, J. Brandt, J. Friedrich, P. Gasteiger, C. Jochim and W. Schubert, Angew. Chem. Intn. Ed., 18, 111 (1979).
27. V.I. Sokolov, Zh. Str. Khim., 15, 747 (1974).
28. Z. Slanina, Contemporary theory of chemical isomerism, Academia, Praha, 1986.
29. C.E. Peishoff and W.L. Jorgensen, J. Org. Chem., 48, 1970 (1983).
30. D. Bonchev, V. Kamenska and D. Kamenski, Monatsh. Chem., 108, 477 (1977); 109, 551 (1978); 110, 607 (1977).
31. I. Ugi et al., Chimia, 421 (1986).
32. A. Panaye, J.A. MacPhee and J.-E. Dubois, Tetrahedron, 36, 759 (1980).
33. V.I. Minkin, L.P. Olekhnovich, Zhdanov Yu. A. Molekulyarny dizain tautomernykh sistem (Molecular Design of Tautomeric Systems), Rostov-on-Don, RGU Publ., 1977.
34. J.G. Nourse, J. Amer. Chem. Soc., 102, 4883 (1980).
35. J. Senior, J. Org. Chem., 3, 1 (1938).
36. Zhdanov Yu.A., Gomologiya v organicheskoy khimii (Homology in Organic Chemistry), MGU Publ., 1950.
37. R. Grimes, Carboranes, New York, Academic Press, 1970.
38. H. Hogeween and P.W. Kwant, J. Amer. Chem. Soc., 96, 2208 (1974).
39. C.M. Giandomenico, L.H. Hanau, S.J. Lippard, Organometallics, 1, 145 (1982).
40. R. Hoffmann, Angew. Chem. Intn. Ed., 21, 711 (1982).
41. K. Fukui, Ibid., 21, 801 (1982).
42. N.S. Zefirov, S.S. Trach, O.S. Chizhov, Karkassniye i politsiklicheskiye soyedineniya (Skeletal and Polycyclic Compounds), Itogi nauki i tekhniki, Orgkhimiya, v. 3, VINITI, 1979.
43. T. Fukunaga, H.E. Simmons, J.J. Wendoloski and M.D. Gordon, J. Amer. Chem. Soc., 105, 2729 (1983).
44. P. Heimbach, H. Schenkluhn and T. Bartik, Pure and Appl. Chem., 53, 2429 (1981).
45. I. Langmuir, J. Amer. Chem. Soc., 41, 868, 1543 (1919).
46. J.E. Ellis, J. Chem. Educ., 53, 1 (1976).
47. M. Elian, M.M. Chen, D.M.P. Mingos and R. Hoffmann, Inorg. Chem., 15, 1148 (1976).

48. M.J.S. Dewar, The Molecular Orbital Theory of Organic Chemistry, McGraw-Hill, New York, 1969.
49. C.K. Ingold, Structure and Mechanism in Organic Chemistry, 2nd ed., Cornell University Press, Ithaca, 1969.
50. K.U. Ingold and B.P. Roberts, Free-Radical Substitution Reactions, Wiley-Interscience, 1971.
51. J. Mathieu, A. Allais and J. Valls, Angew. Chem., 72, 71 (1960).
52. R.B. Guthrie, J. Org. Chem., 40, 402 (1975).
53. W.A. Pryor, J.H. Coco and K.N. Houk, J. Amer. Chem. Soc., 96, 5591 (1974).
54. V.K. Piotrovsky, S.I. Bobrovsky, V.I. Rozenberg, Bundel' Yu.G. and Reutov O.A., Izv. AN SSSR, Ser. Khim., 2285 (1975).
55. O.A. Reutov, I.P. Beletskaya, G.A. Artamkina and A.N. Kashin, Reaktsii metalloorganicheskikh soyedineniy kak redoks-protsessy (Reactions of Organometallic Compounds as Redox Processes), Moscow, Nauka, 1983.
56. J.F. Bunnett, Accts. Chem. Res., 11, 413 (1978).
57. I.P. Beletskaya and V.N. Drozd, Usp. Khim., 48, 793 (1979).
58. L. Eberson, Adv. Phys. Org. Chem., 18, 252 (1982).
59. J.K. Kochi, Pure Appl. Chem., 52, 571 (1980).
60. T.L. Gilchrist and R.C. Storr, Organic Reactions and Orbital Symmetry, Cambridge University Press, 1972.
61. V.A. Nikanorov, 2nd All-Union Conf.on Organometal. Chemistry. Abstr. of papers, p. 6, Gorky, 1982.
62. K.P. Butin, A.I. Ioffe and V.A. Nikanorov, Vestnik MGU, Ser. 2, Khimiy 24, No. 2, 107 (1983).
63. V.A. Uspensky, Usp. Matem. Nauk, 29, 175 (1974); Teorema Goedelya o nepolnote (Goedel's Incompleteness Theorem), Moscow, Nauka, 1982.
64. Yu.A. Vasilyev, V.S. Bagotsky, Usp. Khim., 44, 1953 (1975).
65. A.P. Rudenko, Zh. Fiz. Khim., 57, 1597, 2641 (1983).
66. J. Schnakenberg (Ed.), Thermodynamic Network Analysis of Biological Systems. 2nd ed., Springer Verlag, Berlin, 1981.

Chapter 5

DARC/PELCO METHOD: A TOPOLOGICAL TOOL FOR QSAR SEARCH AND ITS RELIABLE PREDICTIVE CAPABILITY

Christiane Mercier*, Yves Sobel and Jacques-Emile Dubois

(Institut de Topologie et de Dynamique des Systèmes de l'Université Paris VII, 1, rue Guy de la Brosse, 75005 Paris–France)

Introduction

Quantitative Structure-Activity Relationship (QSAR) searches are a useful approach for predicting drug activity, understanding their action mechanism and planning experiments towards optimizing the two first-mentioned purposes. Among these three goals, Prediction is essential because Interpretation and Experiment

Planning are mostly based on underlying prediction capabilities. Predicted values and tendencies are directly used to orient Planning; and the predictive power which conditions the generalization power of the relationship, influences Interpretation. The use of Structure-Activity Relationships in Prediction, generally an ill-defined concept, is greatly hindered by its poor reliability. This notion of reliability cannot be judged by the statistical criteria of the performed relationships when, as is usually the case, these criteria take into account only the data analysis method used and neglect the effective multidimensional variation area of the variables. Our purpose here is to bring out the possibilities of the topological DARC/PELCO method in evaluating and optimizing prediction reliability.

The PELCO method (1) (Perturbation of Environments which are Limited, Concentric and Ordered), proposed in 1967, is one of the correlation methods of the DARC System (2) (Description, Acquisition, Retrieval, Computer aided design). It is quite different from other QSAR methods in its choice of the structural variable, pertaining to ordered topological sites. This very general yet refined description enables it to generate (3) all the types of structural description used in the other methods: groups (5), connectivity indices (6) or general external parameters (7). This method has been significantly applied in such widely diverse fields as reactivity (4), thermodynamic properties (8), spectral properties (9) and biological activities (10). Its success in rationalizing the planning of experiments in the search for more active and less toxic compounds in anticholinergic series (11) showed the efficiency of the DARC/PELCO methodology.

In this paper, we detail the steps needed to obtain a SAR. Beyond the choice of an activity-representing variable, which is not discussed here, SAR search requires the choice of a structural variable well-suited to prediction and the choice of a precise relationship search method. We show the impact of the DARC/PELCO choices in evaluating and optimizing prediction reliability.

This presentation is realized through the study of glucuronic acid conjugation of aliphatic alcohols (12), previously studied by Hansch et al. (13). In this population, the environment contains no heteroatoms, thus clearly showing the influence of topology when connectivity is important. The application to the complete data set is presented in a paper on the use of the DARC/PELCO method by an interactive graphic procedure (3). The example treated herein involves a sample of the complete experimental population, thus bringing out the basic elements of our method.

5.1 Choosing a Structural Variable Well Suited to Prediction: Synchronous Generation of S/HS Space

The importance of the structural variable choice for searching a QSAR is often underestimated. The correlation power, the predictive capability and the scope of the application area are thus limited for lack of a precise and complete account of all the structural modifications.

Two kinds of structural variables can be distinguished in the principal QSAR search methods (10h):

- *Internal structural variables*, derived from purely structural information: groups, as in the Free–Wilson method (5), global connectivity indices (6), ordered sites, such as in the DARC/ PELCO method (1),
- *External structural variables*, usually derived from physico-chemical information measured on external reference populations, such as the electronic, steric and hydrophobic parameters used in Hansch's method (7).

External structural variables have good interpretative power since they derive from experimental considerations. The Hansch structural representation makes it possible to dissociate the hydrophobic, electronic and steric effects of the groups which contribute to activity. However, one can discover no new effects not taken into account by the parameters under consideration. The predictive capability is small because the whole set of parameters is, in most cases, insufficient to represent the structure without ambiguity: the reliable prediction area is difficult to define; the elucidation of structures exhibiting a given activity and the identification of optimal activity structures are hindered. Finally, this representation limits the applicability area to those structures for which the external parameter values can be determined, experimentally or by use of structure – property relationships. It only gives good results for activities where structural influence has the same pattern as in the reference information.

Among *internal structural variables*, connectivity indices often present the same drawbacks without the interpretative power of external structural variables (14). Internal structural variables using groups have a wide applicability area. They are well-suited to activity prediction and to structural elucidation and optimization. However, these methods have no criteria for estimating prediction reliability (10e).

In the DARC/PELCO Structural method, an internal variable pertaining to the ordered generation of an exhaustive topology is used. We show to what extent this variable combines some qualities of other structural variables, avoids some of their drawbacks and permits an approach to prediction reliability.

5.1.1 DARC/PELCO Variable

In the DARC/PELCO method (1), the structural variable makes use of graph theory. It is based on simultaneous representation of all structures (15) whose structural modifications are to be correlated with variations in property and of the population containing these structures. The environment concept is used to describe each structure as an ordered chromatic graph. The principle of synchronous generation of a structure and of the series which contains it, called hyperstructure, is used to describe the population to be studied. This concept and this principle, when applied to an isofocal population, define a multidimensional variable characterizing the chemical structure.

The DARC/PELCO method is applied to a sample population containing 11 of the 30 primary, secondary and tertiary alcohols, tested for their glucuronic acid conjugation capacity (table 5.1). By this example we describe the structures belonging to an isofocal population, the hyperstructure associated with this population, and we define the structural variable characterizing each chemical structure in this population. First, we recall the basic concepts of structure description using an example containing topological

Table 5.1. Glucuronic acid conjugation capacity of aliphatic alcohols in the rabbit

The index MR is the molar percentage of glucuronide formation with respect to the dose of alcohol injected into a rabbit's stomach. The experimental precise on the logarithm of this index is: 0.09 average deviation, 0.27 maximal deviation.

Structure numbers are not continuous because the compounds belong to a larger population (see text). (a) compound yielding no glucuronide, (b) compounds belonging to the retrospective proference.

N°	NAME	ORDERED CHROMATIC GRAPH	log MR Obs.	log MR Calc.
0 [a]	Methanol			
1	Ethanol		−0.30	−0.28
2	Propan-2-ol		1.01	1.02
3	2-Me propan-2-ol		1.39	1.39
4	propan-1-ol		−0.05	−0.03
5 [b]	butan-2-ol		1.16	1.27
6	2-Me butan-2-ol		1.76	1.63
7	iso butanol		0.64	0.69
10	butan-1-ol		0.26	0.22
11 [b]	pentan-2-ol		1.65	1.52
12	2-Me pentan-2-ol		1.75	1.88
13	2-Me butan-1-ol		0.98	0.94
14	3-Me butan-1-ol		0.95	0.94
15	4-Me pentan-2-ol		1.53	1.52

characteristics more varied than those found in a single particular element of the population.

5.1.1.1 *Limited environment concept and structure representation*

Within the framework of the Generation-Description theory (16), each structure is formally assimilated to a graph whose nodes represent non-hydrogen atoms and whose edges represent the bonds between these atoms (Figure 5.1). This graph is chromatic, i.e., coloured by the nature of atoms and bonds. It contains two components:

- the focus FO, constituted by an atom, a bond or a group of atoms characterizing a series of compounds, here the HO–C group of the alcohols,
- the environment E, which brings together the sites (atoms, bonds) outside the focus.

The ordered generation of the environment around the focus is obtained by concentrically propagating an ELCO, Environment which is Limited Concentric and Ordered. Starting from a development origin, which at first is within the focus, an ELCO is constituted of the range of nodes A_i issuing from it and of a range of nodes B_{ij} issuing from A_i. Propagation recurrently and concentrically generates the following ELCOs by taking these B_{ij} as new development origins.

Internal generation inside each ELCO is governed by priority rules taking into account the topological and chromatic criteria intrinsic in each site. The topological criteria, the only ones which intervene in our examples, are essentially reduced to the classification of the A_i sites of one ELCO according to the number of their successors: the most branched, called A_1, is generated first.

During the propagation, each new ELCO takes as its origin of development, the B_{ij} site first generated among those compatible with it. In a few cases the ELCOs generation sequence can no longer be decided by propagation of the already generated order, i.e., several B_{ij} sites are candidates for development origins of several different ELCOs (structure 23 (3)). The number of successors of their development origin intervenes then as a classification criterion.

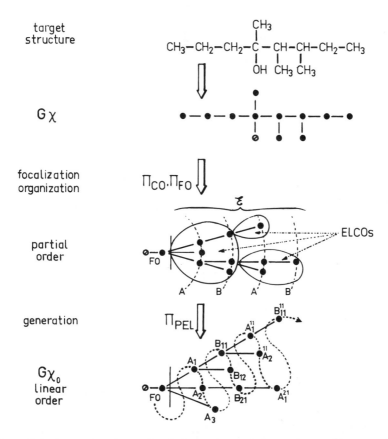

target
structure

G χ

focalization
organization

partial
order

generation

G χ₀
linear
order

Figure 5.1. Generation of an ordered chromatic graph $G_{\chi 0}$ modelling a structure S

The ordered chromatic graph $G_{\chi 0}$, modelling a target structure S, is concentrically organized Π_{CO} around a substructure, called a focus and obtained by the focalization operator Π_{FO}. It is linearly ordered Π_{PEL} by generating the sites of the environment according to strict topological and chromatic criteria, starting from the focus as origin.

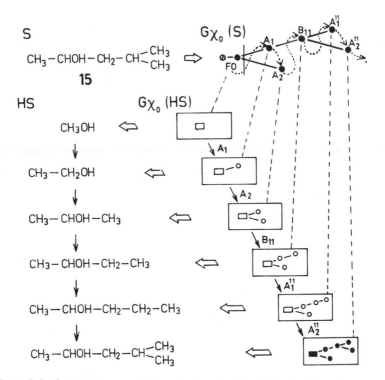

Figure 5.2. Synchronous generation of a structure S and its anteriology hyperstructure HS_{al}

The generation of a structure S, represented by an ordered chromatic graph $G_{x0}(S)$, engenders a series of anteriolog structures (light o) when passing each site from the origin compound associated with the focus to the target structure S (dark •). This series, called an anteriology hyperstructure HS_{al} is itself represented by an ordered chromatic graph $G_{x0}(HS_{al})$.

Should this number be identical, the number of successors of the A_i sites inside each ELCO intervenes. Thus, generation by ELCO propagation induces a linear order over the entire environment E.

To specify that a site belongs to an ELCO having a B_{ij} development origin, its symbol is marked with a superscript $ij : A_2^{11}$ is the site A_2 belonging to the ELCO issued from B_{11}, i.e., whose development origin is site B_{11} of the preceeding layer. The experimental population structures are presented in table 1 as simplified ordered chromatic graphs, in which the geometrical arrangement of the atoms reflects the order induced by the generation over the environment sites. The generation law orders these structures in relation to each other.

5.1.1.2 *"Structure-Hyperstructure" synchronous generation principle and population representation space*

Structure generation is a continuous sequence of steps following a unique path. When passing each site, it generates a structure localized straightway inside the series engendered from the origin compound to the target compound (Figure 5.2). The structures of

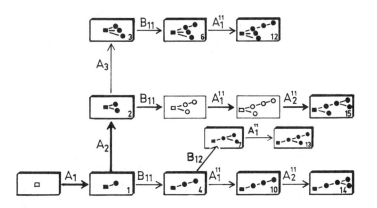

Figure 5.3. Anteriology Hyperstructure HS_{al} associated with the sample population

The synchronous generation S/HS of the 11 experimental structures belonging to the sample population yields to a tree-like anteriology hyperstructure HS_{al} by grouping together the generation paths of each structure.

The same conventions of representing experimental structures in dark ● and affiliated structures in light ○ are adopted in the succeeding figures.

experimental structures,

affiliated structures,

⟶ generation pathway of structure **15**,
⟶ generation pathway of other experimental structures.

this generation pathway are the anteriologs of their successors and constitute an organized population, called an *anteriology hyper-structure HS_{al}* (17).

Every structure population *P* has its corresponding anteriology hyperstructure, obtained by joining the generation pathways of each compound as a tree (Figure 5.3). The set of all anteriology (*al*) and anteriomorphism (am) filiations, created by adjunction of ordered sites existing in the structures of *P*, leads to a network-like *formal anteriomorphism hyperstructure HS(P_{Am}, am)*. *P_{Am}* represents the set of experimental and formal structures related to

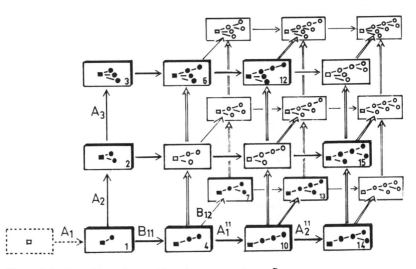

Figure 5.4. Formal anteriomorphism Hyperstructure HS(P̃, am) and DARC Representation space of the sample population

The set of structures of P̃, which are engendered by using the formal relationships am of site adjunction, used to generate the experimental structures of P, are localized in the formal anteriomorphism hyperstructure HS(P̃, am). During the S/HS generation of these structures, the constraint of always having site A_1 is taken into account and eliminates compound **0** associated with the description focus which differs here from the origin of the population ethanol **1**.

——————— anteriology
═══════════ anteriomorphism

those of P. A hyperstructure is represented by a graph. Each node represents a studied or affiliated structure. Each edge represents the topochromatic variation between two structures: the adjunction of site A_1 or A_2 ... or A_2^{11} used to pass from one structure to the other (Figure 5.4). Additional conditions reflect minimal site existence inside the experimental population P and set the boundaries of a $\tilde{P}$ population inside P_{A_m}. They are taken into account during the S/HS generation of $HS(\tilde{P}, am)$ or simply HS hyperstructure representing the actual space of the population studied (10f, 15). The structural zone thus excluded uses the same generation operators as the experimental population. However, its structures do not respect some common structural characteristics which are essential for a homogeneous definition of the population. Here, all 11 experimental structures have site A_1, since compound o, associated with the description focus, has no value for the property studied, not being glycuruconjugated (table 5.1). The limitation of always having site A_1 eliminates methanol o from the formal anteriomorphism hyperstructure and brings back the generation origin to ethanol 1. Prediction search is conducted within this ordered space $HS(\tilde{P}, am)$ of related structures which are generated starting from experimental structures of P, and which are screened for the minimal conditions limiting site existence in order to avoid any extrapolation. Prediction concerning new structures $(\tilde{P} - P)$ is thus achieved by interpolation from the experimental structures. They constitute the so-called "Preference".

5.1.1.3 *DARC/PELCO vectorial structural variable*

Hyperstructure *HS*, which can be considered as a Euclidian vectorial space, provides two tools: the population trace characterizing the representation space of the structures and the topochromatic vector characterizing every structure within this space.

The population trace TR(P) is a graph that groups all the ordered sites appearing at least once in the environments of all the structures in P and used to generate the S/HS space; it provides a basis for representing this vectorial space. This graph can be obtained very simply by superposing all the experimental structures, placing identical sites in corresponding positions (figure 5.5).

The topochromatic vector $\overrightarrow{T}(E)$ is a Boolean vector associated

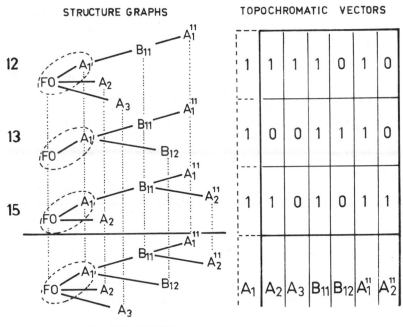

STRUCTURE GRAPHS   TOPOCHROMATIC VECTORS

POPULATION TRACE

Figure 5.5. Simple population trace and topochromatic vectors

In the vectorial S/HS space, whose base representation is called the population trace, each structure graph is represented by a topochromatic vector which indicates the sites used for its generation. In this population, the constraint of always having site A_1 eliminates this site from the description.

with the environment of each structure that indicates the presence of absence of each ordered site. Its dimension is equal to the number of ordered sites used to generate all the structures S and the hyperstructure HS, i.e., equal to the S/HS space dimension. Each component $\zeta(s)$ of this vector takes the value 1 or 0 depending on whether or not the site s has been used for generating the graph of the structure considered (Figure 5.5).

The topochromatic vector provides a quantitative expression of a compound's structure in a population characterized by its trace.

It constitutes an exhaustive vectorial structural variable which directly accounts for the overall topology and chromatism. It includes not only the nature of bonds and atoms, but also the geometrical and stereochemical data (18, 19) of the structure described. This variable is biunivocal and explicit and provides a very simple procedure for retrieving the structure of the compound described (Figure 5.6). Furthermore, it enables one to localize the structure in the $TR(P)$-based vectorial space with respect to the experimental structures (P) and to all the related analogous structures $(\tilde{P} - P)$ (Figure 5.4).

Thus, using the DARC/PELCO vectorial structural variable, a local description is provided for each structure S and a global description localizes each structure with respect to the others

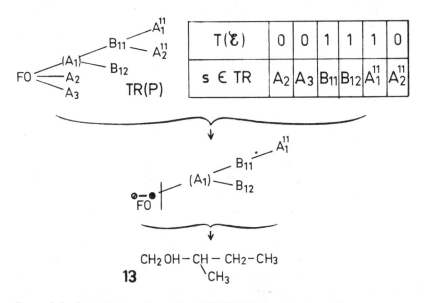

Figure 5.6. Explicit character of the DARC/PELCO structural variable

The topochromatic vector associated with a compound and the trace of the population to which it belongs, make it possible to retrieve the structure of the compound described by a very simple procedure.

within the ordered space *HS* generated from the experimental population.

The topochromatic vectors constitute the basic structural variable used in the DARC/PELCO correlation search method. This variable can be modified by extending or diminishing the dimensions of the structure representation space *S/HS*. The space is modulated according to structural criteria or to criteria pertaining to the information evolving with respect to the structural modifications.

The modulated topochromatic vector $\vec{T}(E)$ which results is obtained by a heuristic modulation of the starting model. The primary topochromatic sites s_p express the smallest formal modifications envisaged for structure generation and constitute the basic elements of this model. More complex sites, resulting from a combination of primary sites, are generated automatically or interactively. Three kinds of complex sites are distinguished:

- *condensed sites* s_c, grouping on a single topological site all the components of some substituents – phenyl, CF_3 or MeO – which always appear simultaneously and are considered as new chromatisms completing the atom chromatism table (10f, 10g),
- *cosites or interaction sites* s_i, designating the simultaneous existence of two or more primary sites, each of which could exist without the others (20),
- *equivalence sites* s_e, grouping those sites with equal influence on property and belonging to homogeneous series of sites, according to structural or physicochemical criteria (21).

Modulated topochromatic vectors involve primary sites, grouped or not in equivalence sites, and also condensed sites, cosites and complex equivalence sites. Condensed sites s_c are used to take account of the eventual gaps in the population. Introducing these sites in the description leads to a new *S/HS* generation in order to define the structural area associated with the experimental population where predictions will be conducted (10f). Interaction sites s_i and equivalence sites s_e are used to search for an optimal Structure-Activity relationship: s_i account for some additivity deviations while s_e reflect regularities in information evolving. Modulated topochromatic vectors constitute a DARC/PELCO structural variable if the modifications of the model dimensions are deduced from a structural analysis of the population or from hypotheses derived from the treatment. They may generate a set of fragmentary variables, ordered or not, or of global variables, if these modifications are made a priori (3).

Thus, the DARC/PELCO structural variable is not only biunivocal, exhaustive, explicit and without any a priori with regard to each structure representation and to relationships between predicted and experimental structures but also flexible and adaptable to any structural and experimental area. An analysis of the criteria that a structural variable must verify to be useful in Drug Design prediction reveals the importance of these qualities.

5.1.2 Suitability of the Structural Variable to Prediction

The use of Structure-Activity Relationship as a prediction tool is rather delicate, as the relationship must be not only precise but also general, powerful and reliable. These qualities essentially depend on the structural variable.

The relationship's precision must be in agreement with the experimental one. Indeed, too great a precision is often illusory and can account for fluctuations arising from experimental errors.

Obtaining a precise relationship necessitates a structural variable being at once non-ambiguous, exhaustive and without any a priori if it is to reveal all the structural modifications influencing activity. Indeed, an ambiguous variable might show the same representation to two structures with different activities, and thus avoid following all of the activity development. A non-exhaustive variable, such as an external variable, makes it impossible to bring out the influence of a structural aspect not taken into account. Thus, applying the extra-thermodynamic HANSCH model often requires introducing *"dummy"* parameters (22) to reflect the influence of some structural modifications not provided for by the hydrophobic, electronic or steric parameters. An a priori variable, which assigns a relative influence to several structural elements, does not enable one to bring out their influence on the biological activity studied, when this latter is not proportionate. Thus, the use of the Molecular Connectivity method for a population of antimicrobial halogenated phenols, leads the authors (23) to suppose that the halogen influence on this actvity is negligible. In fact, the connectivity index $^1\chi^v$, taking chromaticity into account, attributes to the bromine a structural contribution twice that of chlorine, whereas its contribution to antimicrobial activity is approximately the same (10e). The poorer correlation obtained when using $^1\chi^v$ indices rather than $^1\chi$, does not mean that the halogen contribution is negligible. It simply shows that the $^1\chi^v$ scale does not permit taking into account the chlorine and bromine contributions simultaneously (14).

Obtaining a precision in agreement with experimental precision furthermore requires the structural variable to be flexible and adaptable, so as not to attribute to a structural feature fluctuations due to experimental uncertainty. All structural information must be retained, because no universal model exists for the evolving of biological activity. One must determine which model best suits the population studied. However, the model used must not be too restrictive and must make it possible to eliminate those structural elements with no influence on property, or to group elements with the same influence, given experimental uncertainty. This optimization of structural representation permits one to detect regularities in the activity variation with respect to structural modifications, intrinsic to the population studied, and to obtain a simple relationship whose precision agrees with experimental data.

The DARC/PELCO variable which is non-ambiguous, exhaustive and has no a priori for each structure representation is well-suited to obtaining a precise relationship: its flexible and adaptable character makes it useful in any structural and experimental field, especially that of biological activity measurements, seldom precise.

The generality of the relationship, makes it applicable to the two main types of prediction: predicting activity of a given structure and elucidating those structures with a given activity, in particular an optimal activity. Any structural variable permits one to formally predict activity, since prediction in this case follows the same direction as the Structure-Activity relationship. Elucidating structures for a given activity, on the contrary, means building the structure starting from the structural information retained. The structural variable must not only be non-ambiguous and exhaustive but must remain so until the relationship is obtained. Indeed several structures might exhibit the appropriate values for the selected parameters. Thus, for the antimicrobial halogenated phenols example (14), the structural variable selected, the connectivity index $^{1}_{x}$, is ambiguous and not exhaustive. A large number of structures are associated with a single value of this index. The structural variable must furthermore be explicit, i.e., make it possible to build the structure starting from its description. The numbered variable used in the molecular connectivity method is not explicit: even when the set of indices associated with a structure characterizes it without ambiguity, it is practically impossible to retrieve the structural diagram with this index list alone.

The DARC/PELCO variable remains non ambiguous and exhaustive throughout the treatment and is explicit: it makes it possible to retrieve, by a very simple procedure, the structure of the compound whose activity is predicted equal to a required value. Thus, like the structural variable used in the Free–Wilson method, the DARC/PELCO variable is capable of all types of prediction.

The power of the prediction tool reflects the extent of its predictive capacity. Since in topological representations the molecule is divided into atoms and bonds, topology leads to a greater number of predictions than does a division into fragments. Thus, when applied to the antimicrobial halogenated phenols, the DARC/ PELCO topological model leads to 3614 predictions, whereas the Free–Wilson fragmentary model leads to 1254 predictions (10e). The Free–Wilson model must be modified to be applied to the aliphatic alcohol population studied here, as several substituents might exist from a same origin. The modification, generally adopted by the users of this model, assumes that the contributions of identical substituents issued from the same origin are equivalent (24). This a priori hypothesis leads to a non-ordered fragmentary model which cannot be directly compared to the ordered topological DARC/PELCO model. However, the Free–Wilson model can also be modified into an ordered fragmentary model by introducing an order on the substituents. If we apply the two ordered models to the aliphatic alcohol population, we see again that topology, for a same given hypothesis, has a greater predictive capacity. When applied to the 11 alcohol sample population the DARC/PELCO model leads to 10 predictions and the modified Free–Wilson model to 6 predictions. When applied to the entire population (3a), the difference is even greater: the DARC/PELCO predictive capacity results in 338 predictions, whereas the modified Free–Wilson one results in only 91 predictions.

The reliability of the prediction tool reflects its ability to define the boundaries of that set of structures for which reliable predictions can be made. There are structures for which Structure-Activity relationships provide formal prediction, yet to which a precise confidence cannot be accorded:
- either because the predicted structure is too far away from the reference structural area,
- or because the evolving of activity with the structural modifications is irregular.

To estimate prediction reliability, one requires a structural variable by which to define the relationships between predicted and experimental structures, so as to isolate all those containing only the structural modifications defined by the population treated. These relationships must be explicit in order to define the structural elements tested and to characterize the structural area where the relationship can be applied. They must be exhaustive so as to define an interpolation concept and to introduce validity criteria on the prediction structural area. External structural variables are ill-adapted to prediction reliability estimation, because the formal relationships between structures are neither explicit nor exhaustive. Hammet's parameter, for example, implicitly enables one to express some structural relationships, since a σ scale expresses a kind of interstructural relation between the substituents. But these relations are not explicit, since they do not characterize the structural area generated from the structural elements existing in the population treated. They are not exhaustive, since many other interstructural relations might exist. They do not permit validity criteria on the prediction structural area. Hansch's method implicitly considers an interpolation principle on external comportment, insofar as it recommends choosing substituents for an experiment design, so as to maximize the variation range for each type of parameter (25). But these interpolations on external structural parameters do not guarantee interpolation of a given structure on the internal structural level. Predicted structures must contain only those structural modifications defined in the population treated, if the predictions are to be reliable. Indeed, predicting compounds with substituents other than those in the population studied while using relations established on the basis of one-dimensional parameters π, σ, E_s, seems risky even when parameter values are within the variation limits. The predictions are then extrapolations; they correspond to new fields of exploration but go beyond the prediction area (26).

The DARC/PELCO variable both exhaustive and explicit for relationships between predictable and experimental structures, is well-suited to prediction reliability estimation by direct interpolation on the internal structural variable. This structural interpolation is no longer carried out along an axis but in an ordered multidimensional space: the hyperstructure *HS* associated with the population treated. Within this space, structures to be predicted are localized with respect to the experimental ones, and the relations explicitly and exhaustively express the structural variations. This representation makes it possible both to define a structural interpolation concept and to isolate all the structures with only those structural modifications defined in the population treated.

Thus, the DARC/PELCO structural variable answers all the criteria of a prediction tool which is at once precise, general, powerful and reliable. It is adapted for elaborating a simple relationship whose precision agrees with experimentation, and which can provide all types of predictions. Moreover, it is well-suited to prediction reliability estimation by structural interpolation.

5.2 SAR Search: Structural Interpolation in S/HS/I Space

In the DARC/PELCO method (1), comportment evolving is evaluated according to structural modifications – adjunctions of ordered sites – through *S/HS/I* synchronous generation. The relationship which expresses this synchronism is called Topo-Information, because the structural variable is topochromatic, and the search procedure is valid for any associated information. It is obtained by structural interpolation in *S/HS/I* space and its optimization by a heuristic modulation of this space.

5.2.1 Synchronous Generation of S/HS/I Space

Just as the DARC/PELCO vectorial structural variable is based on the synchronous generation of structures *S* and the associated hyperstructure *HS*, so, to determine the relationship, the search method (1) is based on the synchronous generation of information *I* with structures *S* and hyperstructure *HS*.

5.2.1.1 *"Structure-Hyperstructure-Information" synchronous generation principle*

In Topology-Information theory (1, 27), the information concerning a compound is generated by successive contributions of information. They express the perturbations attendant on the progressive introduction of new sites while the compound's ELCO is generated. Thus, the information is constructed parallel to and synchronous with the structure *S* and its hyperstructure *HS*.

During the *S/HS/I* generation of an entity, a structure S_k of an organized population and a related item of information I_k are generated throughout the adjunction of each site s_k. The structural variation s_k corresponding to the generation $S_{k-1} \to S_k$ is associated with an information variation corresponding to the generation $I_{k-1} \to I_k$, called an information perturbation: $p(s_k) = I_k - I_{k-1}$. The information is generated by successively introducing information perturbations $p(A_2), p(A_3) \ldots$. This generation provokes the valuation of the hyperstructure edges by the information perturbations (Figure 5.7).

5.2.1.2 *Topo-Information relationship*

The Topo-Information relationship, which analytically expresses the "Structure-Hyperstructure-Information" synchronous generation principle, uses two vectorial variables: the topochromatic vector $\vec{T}(E)$ whose components $\xi(s)$ characterize the structural environment of a compound and an associated information vector $\vec{I}$ whose components $p(s)$ represent the information perturbations associated with the introduction of sites s in the

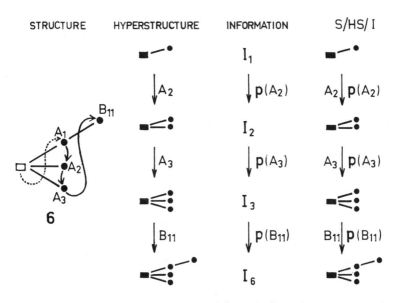

Figure 5.7. "Structure — Hyperstructure — Information" synchronous generation

For each compound, the generation orders the sites of its structure S. It engenders, when passing each site, an organized series of compounds constituting its hyperstructure HS and generates information I by introducing successive information perturbations $p(A_2)$, $p(A_3)$, $p(B_{11})$ associated with each new site introduction A_2, A_3, B_{11}. This generation provokes the valuation of the hyperstructure edges by information perturbations. Here, the generation origin is compound FO-A_1 and the perturbation $p(A_1)$ cannot be evaluated.

environment. The information $I(E)$ associated with the environment within a compound is equal to the scalar product of these two vectors over all sites s included in the population trace:

$$I(E) = \langle \vec{T}(E) \mid \vec{I} \rangle_{TR} = \sum_{s \in TR} p(s) \cdot \xi(s) \qquad (1)$$

where:
- $p(s)$ is the information perturbation associated with the introduction of site s into the environment,

- $\xi(s)$ is the component corresponding to site s in the topochromatic vector $\overrightarrow{T}(E)$.

Information I related to the comportment of a structure belonging to an isofocal population is equal to the sum of information I_0 corresponding to the reference structure associated with the focus of the population and of information $I(E)$ associated with its environment:

$$I = I_0 + I(E) = I_0 + \sum_{s \in TR} p(s) \cdot \xi(s) \tag{2}$$

5.2.2 Regularity and Structural Interpolation

The DARC/PELCO method (1) estimates an average information vector $\overrightarrow{I}(m)$ through use of the topochramatic vectors $\overrightarrow{T}(E)$ and information I associated with each experimental structure of a population P including m entities. The equation is then:

$$I(E) = \langle \overrightarrow{T}(E) | \overrightarrow{I}(m) \rangle_{TR} = \sum_{s \in TR} \bar{p}(s) \cdot \xi(s) \tag{3}$$

where the components $\bar{p}(s)$ of the average information vector express the average information perturbations related to the introduction of sites s into the environment.

The meaning of the average perturbation $\bar{p}(s)$ and its reliability are introduced and explained starting from a local evaluation before its global estimation is presented.

5.2.2.1 *Meaning and reliability of the average perturbation $\bar{p}(s)$*

In a given compound, each component of the information vector $\overrightarrow{I}$ expresses the perturbation $p(s)$ related to the introduction of a site s in its environment. This latter depends on the generation pathway followed to reach this site, globally characterized by its anterior environment $E_A(s)$ which groups all the sites preceding s in the generation law. In a population, a site s is reached by different generation pathways, characterized by different anterior environments $E_A^k(s)$ and the related perturbations $p(s, E_A^k)$ are not the same (Figure 5.8).

In order to treat the entire set of compounds of the hyperstructure generated from the experimental population, one can consider associating with each 'site s a different perturbation term $p(s, E_A^k)$ specific to each anterior environment $E_A^k(s)$. This exhaustive search allows one to obtain a precise value $p(s, E_A^k)$ when $E_A^k(s)$ is the anterior environment of s in an experimentally available structure. It does not allow one to directly estimate the predicted value $p(s, E_A^x)$ when $E_A^x(s)$ belongs to an unavailable structure. We have sought to analyze the very nature of the perturbation term so as to conceive a strategy for acquiring $p(s)$ values allowing for the broadest and most reliable predictions possible.

Each perturbation can be broken down into two parts:

$$p(s, E_A^k) = p(s, E_A^m) + p(s * (E_A^k - E_A^m))$$

where:
- $p(s, E_A^m)$ is a fixed contribution which is always taken into account, depending only on the site itself and linked to its minimal anterior environment $E_A^m(s)$, i.e., the sites conditioning its existence,
- $p(s * (E_A^k - E_A^m))$ is a variable part which reflects the contributions of cosites (5.1.1.3); cosites involved here express simultaneity of site s with sites of E_A^k other than those of E_A^m.

Usually, the second term is small and the various specific perturbations related to a single site are similar given the experimental precision. If this regularity is observed, the perturbation related to the introduction of site s is practically independent of its anterior environment. The generality of this observation leads us to adopt the hypothesis of local average perturbation $\bar{p}_1(s)$. This results in estimating $p(s, E_A^x)$ by the information perturbation $\bar{p}_1(s)$ related to the average anterior environment $\bar{E}_A(s)$ of site s in the population (Figure 5.8):

$$p(s, E_A^x) = p(s, \bar{E}_A) = \bar{p}_1(s) = 1/n \sum_k p(s, E_A^k) \simeq p(s, E_A^m)$$

This hypothesis leads to a simpler and more general Topo-Information relationship. The choice of an average anterior environment $\bar{E}_A$ rather than a minimal anterior environment E_A^m, as reference for estimating the perturbation term, is dictated by criteria of precision, statistic validity and reliability. These qualities

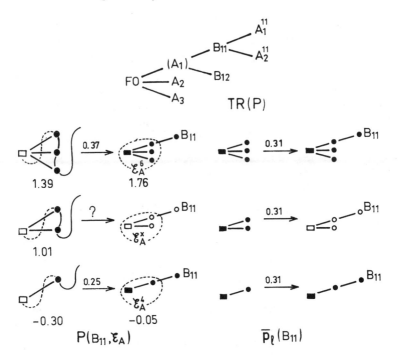

Figure 5.8. Specific perturbation p(s) and local average perturbation $\bar{p}_1(s)$

Within a population, a site s, here B_{11}, is reached by different generation pathways. The anterior environments E_A^4, E_A^6 of this site are different and the associated specific perturbations are not the same: $p(B_{11}, E_A^4) = 0.25$ and $p(B_{11}, E_A^6) = 0.37$. However, considering the experimental error 0.09, these perturbations are similar. Consequently the local average perturbation $\bar{p}_1(B_{11})$ is used for estimating the specific perturbations associated with the environments tested and for predicting the specific perturbations $p(B_{11}, E_A^x)$, which cannot be directly calculated from the experimental compounds.

facilitate the interpretation of the relationship and extend its prediction range. The regularity hypothesis is often valid because too great precision is often illusory. The relationship no longer accounts solely for the phenomenon studied but also for fluctuations arising from experimental errors. However, some irregularities might appear as $p(s * (E_A^k - E_A^m))$ terms bigger than experimental error. A compromise between relationship precision and the extent of its prediction range must then be sought.

The reliability of estimating $p(s, E_A^x)$ by $\bar{p}_1(s)$ depends on localizing E_A^x in relation to the E_A^k tested (Figure 5.9). Reliability is

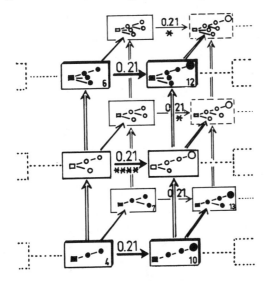

Figure 5.9. Completely or partially reliable estimation of p(s) by the local average perturbation $\bar{p}_1(s)$

The estimation of $p(A_1^{11})$ by the local average perturbation:

$$\bar{p}_1(A_1^{11}) = 1/3(p(A_1^{11}, E_A^{10}) + p(A_1^{11}, E_A^{13}) + p(A_1^{11}, E_A^{12})) = 0.21$$

is completely reliable **** for compounds whose anterior environment lies between those tested because the predictions are based on topological interpolations. This estimation is less reliable * for all the other compounds having A_1^{11} as last site in the generation law (compounds framed in light) because the predictions in this case are based on topological extrapolations.

$\xrightarrow{0.21}$ tested perturbations

$\xrightarrow[****]{0.21}$ interpolated perturbations

$\xrightarrow[*]{0.21}$ extrapolated perturbations

maximal (****) for compounds whose anterior environment lies between those of compounds tested: the predictions then result from topological interpolations (20). This estimation is less reliable (*) for all the other compounds related to site s: the predictions then result from topological extrapolations.

5.2.2.2 *Average perturbations estimation: exploratory correlation*

The procedure for obtaining the local average perturbation $\bar{p}_l(s)$ shows the importance of regularity at the level of a single site. This regularity is better exploited on the whole set of sites

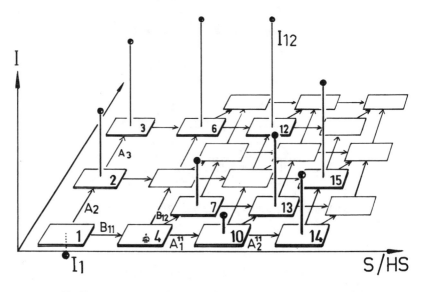

Figure 5.10. Topo-Information SAR search: experimental points in S/HS/I space

The height of each point, which represents the value of the information associated with each structure, is equal to the sum of the reference structure information I_1 and of the perturbations associated with all the site additions leading to its structure. Thus:

$$I_{12} = I_1 + p(A_2) + p(A_3) + p(B_{11}) + p(A_1^{11})$$

Finding the values of these perturbations amounts to seeking a hyperplane of space S/HS/I as close as possible to all the experimental points.

by seeking *global average perturbations* $\bar{p}(s)$.

The experimental compounds are localized in the multidimensional *S/HS/I* space (Figure 5.10). The hyperplane *S/HS* represents the organized space of experimental and predictable structures. The height of each point above this hyperplane represents the value of the information related to the structure S under consideration. It is equal to the sum of the reference structure information and of the

N°	Log MR	TOPOCHROMATIC VECTORS $\vec{T}$ (E)						MODULATED COMPONENTS OF $\vec{\tilde{T}}$ (E)		
1	-0.30	0	0	0	0	0	0	0	0	0
2	1.01	1	0	0	0	0	0	0	0	0
3	1.39	1	1	0	0	0	0	0	0	0
4	-0.05	0	0	1	0	0	0	0	1	0
6	1.76	1	1	1	0	0	0	0	1	0
7	0.64	0	0	1	1	0	0	0	1	1
10	0.26	0	0	1	0	1	0	0	2	0
12	1.75	1	1	1	0	1	0	0	2	0
13	0.98	0	0	1	1	1	0	0	2	1
14	0.95	0	0	1	0	1	1	0	2	1
15	1.53	1	0	1	0	1	1	1	2	0
Sites		A_2	A_3	B_{11}	B_{12}	A_1^{11}	A_2^{11}	$A_2*A_2^{11}$	Σ_d $B_{11}=A_1^{11}$	Σ_r $B_{12}=A_2*\bar{A}_2^{11}$

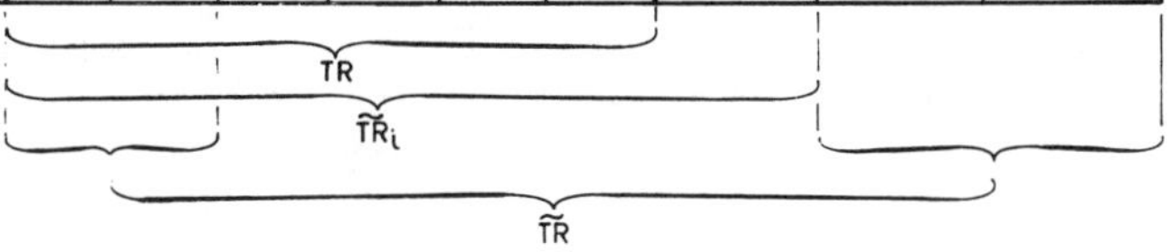

Figure 5.11. Vectorial coordinates in S/HS/I space

The topochromatic vectors $\vec{T}(E)$ and the information associated with each structure are used to establish the exploratory relationship. The modulated topochromatic vectors $\vec{\tilde{T}}(E)$, obtained by a heuristic modulation of the starting model, are used to seek an optimal correlation.

The trace TR groups all the primary sites s_p. The modulated trace $\widetilde{TR}_i$ includes in addition cosites or interaction sites s_i. The modulated trace $\widetilde{TR}$ may include primary sites s_p, interaction sites s_i and equivalence sites s_e.

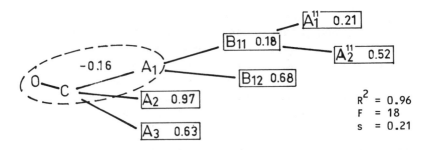

exploratory correlation

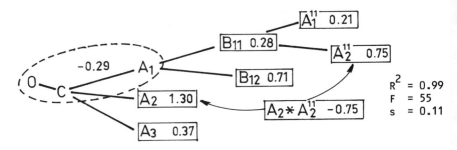

correlation reflecting the additivity deviations

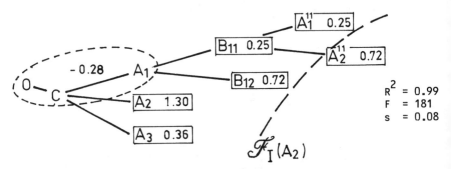

optimal correlation

Figure 5.12. Topo-Information diagrams of the successive correlations

Establishing the optimal Topo-Information relationship involves three successive steps: exploratory correlation, reflecting additivity deviations, detecting regularities. The results of the relationships, the coordinates $\bar{p}(s)$ of the information vector, are perturbations associated with all the site additions leading to structure *S*. Finding the global average values of these perturbations amounts to seeking a hyperplane of space *S/HS/I* as near as possible to all the experimental points. This is accomplished by multiple regression analysis on the vectorial coordinates of the experimental points in *S/HS/I* space. For our sample, this space is a 7 dimensional one (Figure 5.11): 6 topochromatic dimensions correspond to the localization of the points in the *S/HS* hyperplane; one information coordinate corresponds to its height above the hyperplane. The treatment provides the values of the global average perturbations $\bar{p}(s)$ which are the components of the average information vector $\vec{I}(m)$, as well as the statistical criteria for the correlation.

The results of the regression carried out on the sample population of 11 alcohols, with the components of the topochromatic vector as structural parameters, are visualized on a Topo-Information diagram. The coordinates $\bar{p}(s)$ of the information vector are located on the sites s of the experimental population trace (Figure 5.12). The values inside each rectangle represent the perturbations of information associated with the introduction of each site in the generation. For example, the value 0.18 represents the perturbation of information related to the introduction of site B_{11} in a compound's generation. These values do not represent a real measure of the perturbation but an estimation of its value for the entire population. Thus, the real value of the perturbation related to the introduction of a methyl in B_{11} is 0.25 ($-0.05 - (-0.30) = 0.25$) for ethanol, but its estimated value for the set of compounds is 0.18.

---

located on the population trace which is then called a Topo-Information diagram. The equation of the optimal relationship can be written:

$$\log MR = -0.28 + 1.30\,A_2 + 0.36\,A_3 + 0.25\sum_{I} + 0.72\sum_{b}$$

where

$\sum_{I}$ groups the chain-lengthening sites B_{11} and A_1^{11},

$\sum_{b}$ groups the branching sites B_{11} and A_2^{11} (in the absence of A_2)

The exploratory correlation obtained is most often satisfactory but rarely the best. The values of statistical criteria, the square of correlation coefficient R^2, the overall F test and the standard deviation s permit one to judge the credibility of the results (28). Those of the exploratory correlation performed here ($R^2 = 0.96$, $F = 18$, $s = 0.21$) are satisfactory. The square of the correlation coefficient tells us that the regression explains 96% of the total variance. The overall F test indicates that the introduction of parameters, other than that attributed to the focus, is significant with a <1% risk. However, inaccuracy for the calculated values (average deviation: 0.13) is slightly greater than for the observed values (average deviation: 0.09).

Analyzing the exploratory correlation sometimes suggests deviations from the additivity of some site perturbations. It often

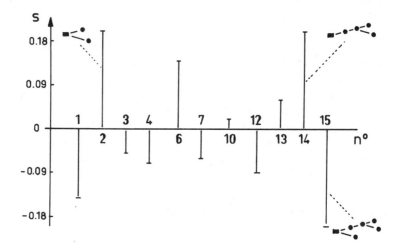

Figure 5.13. Deviations between observed and calculated values in the exploratory correlation

Deviations greater than twice the experimental average deviation are observed for three compounds: +0.20 for minimal compounds **2** and **14** respectively associated with sites A_2 and A_2^{11} and −0.20 for compound **15** including these two sites simultaneously. This suggests that the perturbations associated with sites A_2 and A_2^{11} are not additive.

suggests a greater regularity in the activity variation with respect to the structural modifications. Deviations from additivity are detected by analyzing the important deviations between values observed and calculated during application of the Topo-Information relationship. Thus, deviations greater than twice the experimental average deviation are observed here for three compounds (Figure 5.13): +0.20 for the minimal compounds 2 and 14, respectively associated with sites A_2 and A_2^{11} and -0.20 for compound 15 including simultaneously those two sites. This suggests that the perturbations related to sites A_2 and A_2^{11} are not additive. The regularities are suggested by analyzing the Topo-Information diagram: here, the perturbations related to the chain lengthening sites B_{11} and A_1^{11} are equivalent (0.18 and 0.21), given the experimental precision which is 0.09.

5.2.3 Heuristic Modulation of S/HS/I Space: Optimizing the Topo-information Relationship

Optimization of the Topo-Information relationship is based on the criterion of experimental precision and aims at obtaining a relationship both simple and as precise as the measurements. It includes two successive steps, based on a heuristic modulation of the initial *S/HS/I* space representing the experimental data. These steps aim at best reflecting the additivity deviations detected in the exploratory correlation and the regularities suggested by the next correlation.

The strategy of modulating *S/HS/I* space to reflect the additivity deviations includes extending steps by introducing new structural variables taking into account the non-additivity of some perturbations. When important deviations between observed and calculated values are observed, then the specific perturbations $p(s)$ related to these experimental compounds are too far from the average perturbation $\bar{p}(s)$: simultaneities of sites between site s and the variable part of its anterior environment influence the property. Here the specific perturbations related to sites A_2 and A_2^{11} are different according to their introduction either in their minimal anterior environment or in presence of each other (Figure 5.14). This simultaneity leads to a decrease of about 0.70 in each specific

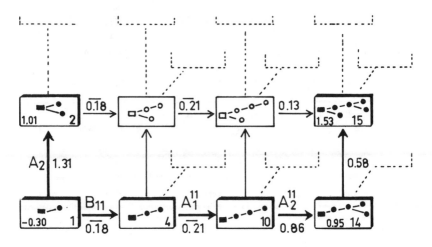

Figure 5.14. Different specific perturbations and detecting cosites

The specific perturbations associated with sites A_2 and A_2^{11} in their minimal anterior environment, respectively 1.31 and 0.86, are about 0.70 less when these sites are present simultaneously. This additivity deviation is taken into account by introducing a cosite or interaction site $A_2^* A_2^{11}$ during the optimization of the relationship.

0.18 : average perturbations obtained in the exploratory corrrelation,
1.31 : specific perturbations.

perturbation. These additivity deviations are taken into account by introducing cosites indicating the simultaneous presence of two or more sites, here the two order cosite $A_2^* A_2^{11}$. The space in which the topochromatic vectors are located by the primary site (s_p) coordinates is consequently extended by introducing cosite or interaction site (s_i) coordinates (Figure 5.11).

The vectorial equation of the Topo-Information relationship is then expressed by the sum of the average perturbations $\bar{p}(s_p)$ associated with the primary sites s_p localized in the trace and by that of the average perturbations $\bar{p}(s_i)$ associated with the interaction sites s_i detected in the environment:

$$I(E) = \sum_{sp \in TR} \bar{p}(s_p) \cdot \xi(s_p) + \sum \bar{p}(s_i) \cdot \xi(s_i) \qquad (4)$$

The first term accounts for strict additivity. The second term expresses the possible binary or higher level interactions. When the set of primary sites and detected interaction sites are grouped in the modulated trace *TR*, this equation becomes similar to equation 1:

$$I(E) = \sum_{s \in TR} \bar{p}(s) \cdot \xi(s) \tag{5}$$

where s is either a primary site s_p or an interaction site s_i.

The regression, carried out with the primary sites and the cosite $A_2 * A_2^{11}$, is excellent. The statistical criteria $R^2 = 0.99$, $F = 55$, $s = 0.11$ correspond to a less than 0.1% risk. This correlation describes with great precision the variations in glucuronic acid conjugation capacity along with structural modifications (average deviation: 0.07). However, the contributions of several sites are equivalent among themselves, granted experimental precision (0.09). This suggests greater regularity, possibly hidden by fluctuations arising from experimental errors.

The strategy of modulating *S/HS/I* space to detect regularities includes both extending and reducing steps. Extending steps introduce new structural variables, redundant but better fitting activity development. Reducing steps eliminate the less significant structural variables, particularly those whose redundance, given experimental inaccuracy, comes from regularity. The redundant structural variables are equivalent sites s_e interactively constructed by grouping sites belonging to homogeneous series of the same type and according to examination of the associated perturbation $\bar{p}(s)$ values (10f) (21). The Topo-Information equation keeps the form of equation 5, with a modulated trace including the equivalence sites. This formal similarity must not hide an important difference: the components $\xi(s_e)$ of the modulated topochromatic vector may take values which are greater than 1 if several equivalence sites are present in the same environment. Thus, the environment of structure 14 (Figure 5.11) includes only one of the two equivalent sites $B_{12} = A_2^{11}$ but both equivalent sites $B_{11} = A_1^{11}$:

$$\xi(B_{12} = A_2^{11}) = 1, \qquad \xi(B_{11} = A_1^{11}) = 2$$

Applied to the glucuronic acid conjugation capacity of aliphatic

alcohols, this strategy leads to two main classes of homogeneous sites:
- chain-lengthening sites B_{11} and A_1^{11} grouped in the Σ_l variable,
- branching sites B_{12} and A_2^{11} (in the absence of A_2) grouped in the Σ_b variable.

Sites A_2 and A_3 remain individualized: they account for the different comportment between alcohols according to the nature of the hydroxyl group (primary, secondary, tertiary).

The heuristic strategy of modulating finally reduces space *S/HS/I* to 5 dimensions corresponding to the activity variable completed by these 4 structural variables A_2, A_3, Σ_l, Σ_b, selected as the most significant. The optimal Topo-Information relationship whose statistical criteria $R^2 = 0.99$, $F = 181$, $s = 0.08$, are excellent (Figure 5.12), describes the glucuronic acid conjugation capacity as precisely as experimentation.

5.3 Drug Design Application

In addition to the tool for searching an optimal Structure-Activity relationship, the DARC/PELCO method also has several tools with which to exploit results in Drug Design. These concern the interpretation of the relation, its predictive capability and experiment planning.

5.3.1 Interpreting the Structure-activity Relationship

This interpretation relies on an analysis of the Topo-Information diagram. The analysis is first done locally to specify the influence of structural modifications on activity, then globally, pointing out one or several frontiers of influence on activity.

Local analysis

Local analysis of the Topo-Information diagram depends on the kinds of variables selected and on the sign and absolute values of the contributions assigned to them. The regular influence of certain structural modifications grouped in equivalence sites makes for a generalization of certain effects – chain-lengthening (3) (14) (21), branching (3) (21), direct substitution on an aromatic nucleus (10f) – which can eventually be interpreted in terms of physico-chemical effects. Exceptional contribution values, either by their sign or by their absolute value, can draw attention to specific effects. Detecting these singularities in behavior can suggest new fields of investigation towards an area of higher activity with an optimized strategy (10d, f). Still others point out specific effects and draw attention to a mechanism that can subsequently be verified experimentally or, if the mechanism is a familiar one, can be quantitatively evaluated with statistical validity.

In this example, local analysis of results shows that the glucuronic acid conjugation capacity depends on three factors:

- *The nature of the hydroxyl group*: secondary and tertiary alcohols are strongly glycuroconjugated, approximately ten times better than that of the isomeric primary alcohols. This is evidenced by a very important contribution of site $A_2(1.30)$ with respect to chain lengthening $\Sigma_l(0.30)$.
- *Branching*: primary alcohols are very sensitive to branching effects which greatly increase their glycuroconjugation capacity. For secondary and tertiary alcohols, a saturation effect appears for β branching: site A_2^{11} has zero influence when combined with A_2.
- *Chain-lengthening*: in all three groups of alcohols, the chain-lengthening increases glycuroconjugation slightly but regularly.

This analysis provides an explicit and more precise interpretation of alcohol metabolism. It is generally stated that alipathic alcohols are metabolized and eliminated from the body in two ways:

- by *in vivo* oxidation and elimination of the products (acids, aldehydes, ketones and CO_2) in urine and in exhaled air,
- by conjugation with glucuronic acid and elimination of glucuronides in urine.

The elimination for a given alcohol depends mainly on the number of carbon atoms, the nature of the hydroxyl group and the extent of the branching of the carbon chain (12). For the authors, the general conjugation order of aliphatic alcohols is tertiary > secondary > primary and, since oxidation goes contrary-wise, they conclude that the unoxidized portion of alcohol is subjected to glycuroconjugation. The results here class secondary alcohols with tertiary ones, which agrees with the fact that ketones can give glucuronides. Moreover, since the steric effect seems more important for primary alcohols and inhibits their oxidation, we can conclude, in agreement with Hansch (13), that steric hindrance increases the glucuroconjugation of primary alcohols by inhibiting their oxidation.

Global Analysis

There are various kinds of structural influence frontiers on activity:

- An *activity frontier* F_A, beyond which site influence on activity is zero or unfavorable (10d). This frontier surrounds the favorable activity environment E_F^* and is of interest when seeking the optimization of a series.
- A *zero influence frontier* F_0, beyond which site influence on activity is zero. This frontier surrounds the active environment (1a), noted E^* and is of interest when a prediction law is sought.
- One or several *inhibiting frontiers* $F_I(s)$, beyond which site influence is inhibited by the presence of site s(10d). The additive character of perturbation terms associated with sites located beyond this frontier is no longer valid in the presence of site s.
- One or several *equivalence frontiers* F_E, beyond which site influence on activity is equivalent.

The population studied here is too limited to detect the activity frontier F_A and above all the zero influence frontier F_0, most interesting for this study whose aim is to find a prediction law. This latter is achieved on the complete example (3). The inhibition frontier $F_I(A_2)$, shown in dots in figure 13, goes beyond site A_1^{11}. It corresponds to a zero influence frontier for secondary and tertiary alcohols.

The strength of these arguments depends on the reliability of the QSAR parameters and on the structural area of reliable validity of the QSAR.

5.3.2 Predictive Capability of the Relationship

The reliable prediction area of the Topo-Information relationship consists of structures whose predicted activity is obtained by interpolation and which constitute proference. Within this area one can distinguish degrees of reliability. In some cases one can use extrapolation without reducing reliability, and this gives rise to pseudoproference. A heuristical system based on the hyperstructure concept automatically enumerates and locates all the structures of the proference or pseudoproference and systematically measures prediction reliability (3).

5.3.2.1 *Proference and evaluation of predictive capacity*

In the DARC/PELCO method, the Topo-Information predictive capacity is measured by Proference PR (1b) including all the structures of $\tilde{P}$ generated from the studied population P and which do not belong to it:

$$PR = \tilde{P} - P$$

The *S/HS* synchronous generation of the structures of the studied population P and of its hyperstructure $HS(\tilde{P}, am)$ is used to locate the predictable structures of PR with respect to the experimental structures of P. The Topo-Information relationship expresses the synchronous generation $S/HS/I$ of associated information. It is used to predict the value of the information I associated with the structures of the proference PR.

In this example proference includes ten structures belonging to the complete population $\tilde{P}$ of 21 structures generated from the sample population P of 11 structures:

$$PR = \tilde{P} - P = 21 - 11 = 10$$

These ten predictable structures are located in the hyperstructure $HS(\tilde{P}, am)$ with respect to the experimental structures. The results of the Topo-Information relationship are used to calculate the predicted values of their activity. We see them on the hyperstructure (Figure 5.15): each average perturbation value $\bar{p}(s)$ is located on the arc representing the adjunction of site s. This figure illustrates the principle of synchronous generation of information together with structures. By following the generation pathway of a structure one gradually generates information by successive additions of different average perturbations $\bar{p}(s)$ associated with the adjunctions of sites used for its generation. Thus, the information associated with structure 11, pentanol-2, is generated from that

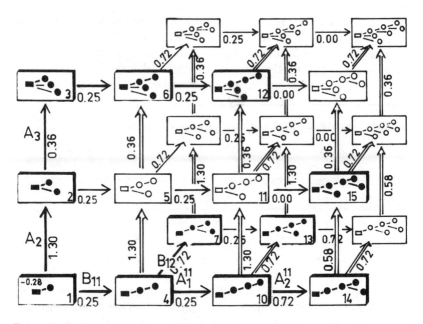

Figure 5.15. Hyperstructure weighted by the results of the Topo-Information relationship

The values of the average perturbations $\bar{p}(s)$, associated with the site additions, are located on the arcs of the hyperstructure $HS(\tilde{P}, am)$. They are used to retrieve the acitivity value of the 11 experimental structures of P (•) and to estimate that of the 10 predictable structures of $\tilde{P} - P$ (o) which constitute the proference. For example, the predicted value of structure **11** is generated from that of ethanol **1**, the origin, by successively adding the different average perturbations $\bar{p}(A_3)$, $\bar{p}(B_{11})$ $\bar{p}(A_1^{11})$, associated with the site adjunctions used to generate its structure:

$$\log MR \text{ predicted} = -0.28 + 1.30 + 0.25 + 0.25 = 1.52$$

of the origin, ethanol, by successive contributions $\bar{p}(A_2)$, $\bar{p}(B_{11})$, $\bar{p}(A_1^{11})$:

$$\log MR \text{ predicted} = -0.28 + 1.30 + 0.25 + 0.25 = 1.52$$

The calculated activity values associated with the reference structures correspond to the regularity degree which is compatible with experimental precision.

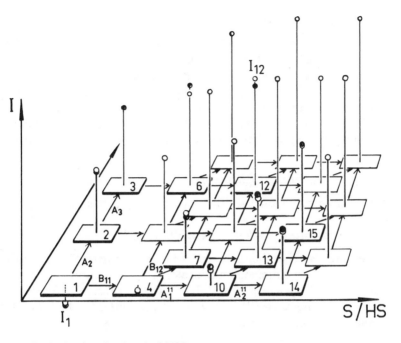

Figure 5.16. Predicted points in S/HS/I space

The white points associated with the retrieved values of the experimental structures in dark outlines, and with the estimated values of the predictable structures
in light outlines, are localized in space S/HS/I with respect to the experimental points
in black.

All the points associated with the predictions, calculated values
of the reference structures and estimated values of the predictable
structures are located in the space *S/HS/I* with regard to the experimental points (Figure 5.16). They are situated in the hyperplane
of the optimal Topo-Information relationship. The set of these
points furnishes a visual display of how activity develops along the
hyperstructure and makes it possible to define the areas of regular
activity variation with structural modification.

Proference covers both retrospective and prospective aspects
     Retrospective proference is equal to the number of experimental
compounds not used for treatment:

here, PR (retrospective) $= 2$

It enables to check the predictive validity of the correlation. The values associated with structures 5 and 11, not included in the treatment and calculated with the help of the established Topo-Information relationship, agree with the observed values, given experimental precision (Table 5.1).

Prospective proference equals the number of compounds belonging to the proference and not belonging to retrospective proference:

here, PR (prospective) $= 10 - 2 = 8$

It expresses quantitatively the extent of the prediction range.

5.3.2.2 *Evaluating predictive reliability*

Predictive reliability is measured by applying two interpolation criteria on a two-dimensional scale. The structural criterion is linked to S/HS synchronism, the regularity criterion to $S/HS/$ synchronism. Proference PR is organized according to the reliability measured on this scale (20).

Structural interpolation

Prediction reliability relies on the structural interpolation concept. As for a classical mathematical function, predictions by interpolation are more reliable than by extrapolation. Moreover, regularity enhances the reliability of interpolated and, especially, extrapolated predictions.

Structural interpolation takes place in a multidimensional space. Each structural dimension refers to a site which may be a primary site s_p or a cosite s_i (interaction site). The cosites augment the number of directions taken into account to evaluate surroundings of each predictable structure by experimental structures in hyperstructure HS.

The value of the information associated with a structure S is obtained by structural interpolation if S contains only sites existing in structures of reference population P. When the prediction is based on a structural interpolation solely at the level of the primary sites, reliability is minimal and is noted F_{min}. When the interpolation takes into account all the sites and cosites, reliability is maximal and is noted F_{max}.

All structures belonging to the proference are predictable with minimal reliability F_{min} since they are generated by adjunction of sites used to generate the structures of P. On the other hand, only the anteriomorph structures (17) of experimental structures are predictable with maximal reliability F_{max}: they are situated on generation pathways linking structures of the population treated. Thus, structures 5 and 11, anteriomorph of experimental structures 6 and 12, are predicted with maximal reliability. Cosite $A_2 * B_{11}$ of structure 5 and cosites $A_2 * B_{11}$ and $A_2 * A_1^{11}$ of structure 11 exist in structures 6 and 12 of population P (Figure 5.17).

Reliability scale

Between these two extremes F_{min} and F_{max}, we define a two dimensional reliability scale. This assigns to the prediction associated with a structure a reliability F_k^j characterized by a level k and a degree j of reliability.

The level k of reliability, linked to *S/HS* synchronism, reflects the extent to which the structure is surrounded by experimental structures in hyperstructure *HS*. It is defined by the complexity level of the sites taken into account for estimating the reliability. Thus on level $k = 1$ only primary sites are considered; on level $k = n$ all 1 to n order sites are considered, i.e., the primary sites and the cosites reflecting the simultaneous existence of 2 to n primary sites (cf.I.1.3).

The degree j of reliability, linked to *S/HS/I* synchronism, introduces nuances into strict structural interpolation when the activity evolving along the hyperstructure is sufficiently regular. It is defined by the reliability degree of the perturbation $p(s)$ estimation associated with each site of the structure. This latter is maximal ****, if the site exists in at least one structure of P. It is *** or ** if the estimation is based on hypotheses deduced from regularity detected by the treatment. It is * if the estimation is based on a priori hypotheses. Prediction associated with structure possesses a reliability of degree j on the level k if perturbations associated with l to k order sites are estimated with a reliability degree at least equal to j.

In this example, the reliability analysis is carried out on level 2. The perturbations associated with primary sites all have a maximal degree of reliability **** since these sites all exist with their minimal anterior environment in P. Reliability degrees of perturbations associated with binary cosites are shown on figure 5.18. Here, all perturbations associated with untested binary cosites are con-

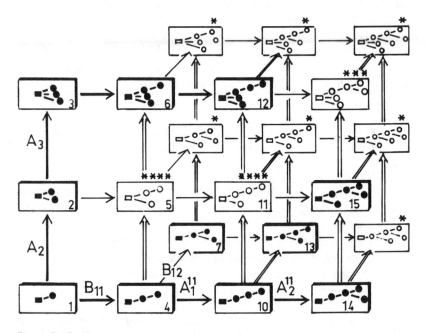

Figure 5.17. Hyperstructure weighted by prediction level 2 reliability

All the new structures $\tilde{P} - P$ (in light ○) engendered starting from the 11 experimental structures of the sample population P (in dark ●) are predictable with a minimal reliability: PRmin = PR = 10. The two anteriomorph structures of P, **5** and **11**, are predictable with a maximal reliability: PRmax = 2. Intermediate reliability is determined at level 2, by examining, in each structure, the binary cosites controlling the reliability:

*   : $A_2 * B_{12}$ and $B_{12} * A_2^{11}$,

**   : no cosite has this reliability in this sample population,

*** : $A_3 * A_2^{11}$

sidered to be zero. The reliability of this estimation is related to the more-or-less justified character of the zero hypothesis. Reliability is *** if the cosites include a zero contribution site; such sites are situated beyond the zero influence frontier F_0 or beyond the inhibiting frontier $F_I(A_2)$ for sites linked to the existence of A_2: cosite $A_3 * A_2^{11}$ is considered as having a zero contribution with *** reliability. Reliability would be ** for cosites including a site linked to the existence of A_2 and a site tested with A_2; this null hypothesis is based on the behavior analogy of secondary and tertiary alcohols. In this sample population no cosite belongs to this category, since all the sites tested with A_2 are also tested with A_3 or situated beyond the inhibiting frontier. Reliability * is attributed to cosites belonging to none of these categories, here, those cosites including site B_{12}.

By considering all the binary cosites existing in the structure, we obtain an automatic determination of the reliability degree on level 2 (figures 5.17 & 5.18):

- structures predicted with **** reliability possess only binary cosites of **** reliability;
- structures predicted with *** reliability possess at least one cosite of *** reliability, in this case cosite $A_3 * A_2^{11}$;
- structures predicted with * reliability possess at least one * reliability cosite, in this case cosites $A_2 * B_{12}$ and $B_{12} * A_1^{11}$.

Organizing preference in terms of reliability

Grouping structures whose activity prediction is carried out with similar reliability creates reliability areas within proference *PR*.

Proference of minimal reliability $PR_{\min}$ covers the whole proference for populations where each site exists with its minimal anterior environment. Here, the 11 experimental structures make it possible to predict with minimal reliability the activity of the 10 proference structures:

$$PR_{\min} = PR = 10$$

Proference of maximal reliability $PR_{\max}$ includes the set of anteriomorphs (17) of the reference population structures: $P_{Am}(P)$. Here, the 11 experimental structures make it possible to estimate

Sites \ Sites	A_2	A_3	B_{11}	B_{12}	A_1^{11}	A_2^{11}
A_2	5	✕	**** 4	*	**** 2	**** 1
A_3		3	**** 2	*	**** 1	***
B_{11}			8	✕	✕	✕
B_{12}				2	**** 1	*
A_1^{11}					3	✕
A_2^{11}						2

Figure 5.18. Reliability degrees of the p̄(s) associated with binary cosites

The numbers on the diagonal represent the occurrence of the primary sites in the sample population, the others the occurrence of the binary cosites. Perturbations p̄(s) associated with binary cosites are reliable as follows:

**** , if the cosite exists in a structure of P,
*** , if the cosite causes the intervention of a site situated beyond the zero influence frontier F_0 or a site situated beyond the inhibiting frontier $F_i(A_2)$ and a site linked to the existence of A_2,
** , if the cosite causes the intervention of site A_3 and a site tested with A_2,
* , if the cosite belongs to none of these categories.

In this example population, all sites tested with A_2 are also tested with A_3 or situated beyond the inhibiting frontier. Thus there is no cosite of ** reliability.

with maximal reliability the predicted activity of only two structures 5 and 11 which are anteriomorph of the experimental structures 6 and 12 (Figure 5.17):

$$PR_{max} = P_{Am}(P) = 2$$

Among the intermediate preferences, only those at level 2, for which the reliability estimation takes into account the binary cosites, are determined here. Structures belonging to these preferences are localized in hyperstructure $HS(\tilde{P}, am)$ generated from the sample population of 11 structures (Figure 5.17).

Here, the preference is developed in five preferences of increasing reliability:

$$PR_{min} = PR_1^{****} = 10 \rightarrow PR_2^{*} = 7 \diagup PR_2^{***} = 1 \diagup PR_2^{****} = 2 \rightarrow PR_{max} = PR_n^{****} = 2$$

More generally, preference PR_k^j (degree j at level k) contains the set of structures whose information value is predicted with a reliability at least equal to F_k^j. Preference at level 1 has always a **** reliability degree because predictions with a lesser degree of reliability belong to what we call the "Pseudoproference".

5.3.2.3 *Prediction by extrapolation: Pseudoproference*

When one can hypothesize a progressive evolving of activity outside the experimental range, the prediction area can be extended to those structures whose prediction is based on extrapolations on the level of primary sites and which constitute pseudoproference (21).

The prediction structural area is constructed from an extension of the S/HS space reflected by an extension of the trace $TR(P)$ of the population treated (Figure 5.19). By assigning values to average perturbations of sites which depart from the population trace we can predict information associated with these structures.

The activity predictions associated with the structures of the pseudoproference are deduced from the regularities detected in the experimental population P. In our example, two regularities are detected for primary sites. Thus the regular contribution of chain-lengthening sites (B_{11} and A_1^{11}) is assigned to two sites of the same type departing from the trace: $\bar{p}(B_{21}) = \bar{p}(B_{11}^{11}) = 0.25$.

Likewise the regular contribution of branching sites (B_{12} and A_2^{11}) is assigned to one more site of the same type: $\bar{p}(B_{12}^{11}) = 0.72$.

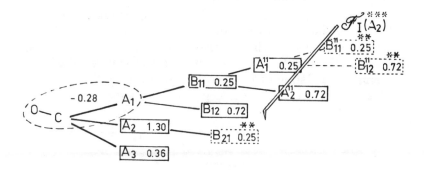

Figure 5.19. Extending regularity areas and predicting by extrapolation

The regularities detected are used to assign a value to the average perturbation of similar sites that depart from the trace of the population treated. Predictions deduced from a first order regularity for primary alcohols have ** reliability; those deduced from zero order regularity for secondary and tertiary alcohols have *** reliability.

[box] sites of trace TR(P)

[dotted box] sites of the extended trace.

Given the existence of an inhibiting frontier $F_I(A_2)$, a zero contribution is assigned to the sites located beyond this frontier, for secondary and tertiary alcohols: $\bar{p}(B_{11}^{11}) = \bar{p}(B_{12}^{11}) = 0$, when A_2 is present.

The extended trace includes, in addition to the sites of *TR(P)*, the sites for which we can estimate the associated perturbations by means of such regularities (figure 19). It engenders a population $\tilde{P}'$ of structures. The pseudoproference *PR* includes all the structures of $\tilde{P}'$ not belonging to the population $\tilde{P}: PR = \tilde{P}' - \tilde{P}$. In this example, the pseudoproference includes 52 structures:

$$PR\psi = 73 - 21 = 52.$$

Reliability of Pseudoproference

When one extends the prediction area, the reliability degree (20) of the new predictions depends on the quality of the regularity hypothesis which permitted this extension outside the known experimental area. This is never as high as **** since the predictions are obtained by extrapolation and not by interpolation. It is limited to * if the regularity hypothesis is a priori posited. It can reach ** or *** if the hypothesis is induced from regularities detected by the treatment.

In fact, the reliability degree of the new predictions is deduced from that of the average perturbations associated with the extended trace sites. A quantitative evaluation of the regularity of activity evolving in *S/HS/I* space assigns the reliability degree ** or *** to each new site. It is ***, if the estimation is deduced from a zero order regularity, i.e., by prolongation of a homogenous series of sites causing a zero perturbation on activity. It is **, if the estimation is deduced from a first order regularity, i.e., by prolongation of a homogenous series of sites causing a same non-zero perturbation (figure 5.19).

For the population treated here, we used only hypotheses inferred from the results:

$$PR_1^{***} = 32 \qquad PR_1^{**} = 20$$

Reliability estimation measures the impact of data choice on structure-activity prediction. It is used to increase reliable prediction in experiment planning.

5.3.3.2 Planning Experiments Towards Prediction Reliability Optimization

Experiment planning aims at achieving a goal at minimal cost given certain constraints. In the case of QSAR search, the goal may be: mechanism elucidation, search for maximal activity and minimal toxicity compounds.... Since planning is always based on prediction, the search is for the most suitable predictive area. Cost is essentially expressed in the number and the difficulty of new syntheses or measurements. Constraints are due to experimental difficulties (synthesis, solubility, availability) or to simultaneous optimization of another property such as toxicity. The result is usually a list of compounds to be synthesized and/or experimented.

In the DARC system, experiment planning is based on the

notion of the key or optimal population to be handled. The list searched for is thus the part of the key population not yet available. The design of this population obeys the criterion of prediction reliability optimization. The diverse experimental criteria are considered for modulating the preliminary choice of compounds realized according to pure reliability criteria.

Before presenting the main stages of the experiment plan, we define the theoretical key population and show the impact of its choice on Proference reliability.

5.3.3.1 *Theoretical key population and completely reliable Proference*

The theoretical key population is the optimal population to be handled if only the prediction reliability criterion is taken into account.

The selection of compounds constituting this population is based on the interpolation hypothesis. When estimating the local average perturbation $\bar{p}_1(s)$ associated with a site s, two limit compounds, associated with the minimal and maximal anterior environments of this site must be used. This choice permits a completely reliable estimation of the specific perturbations $p(s, E_A^x)$ by their local average perturbation $\bar{p}_1(s)_m^M$ in all the compounds which include site s as their last site. Indeed all predictions are thus based on a topological interpolation since the anterior environments of these compounds lie between those tested (Figure 5.20).

The minimal and maximal compounds associated with each site of the trace and with the detected cosites constitute the key population P_c. When the treatment requires only primary or equivalent sites, the resulting theoretical key population P_c^1 is labelled level 1. It groups the minimal and maximal compounds associated with the focus and the topochromatic sites of the trace (the level 1 sites). When the treatment requires introducing level k cosites ($k \geqslant 2$), then the theoretical key population labelled level k also includes the minimal compounds associated with those detected cosites (Figure 21).

When the population treated is the theoretical key population P_c, the maximal reliability proference PR_n^{****} includes all the proference structures.

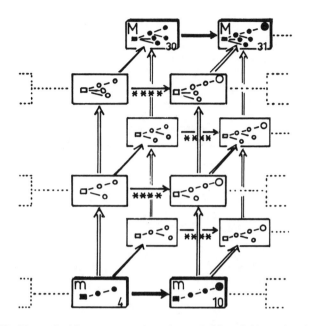

Figure 5.20. Theoretical key compounds and completely reliable estimation of the p(s) by local average pertubation $\bar{p}_1(s)$

The choice of two border compounds, the minimal compound **10** and the maximal compound **31** associated with the minimal and maximal anterior environments E_A^m and E_A^M of site A_1^{11}, makes possible a completely reliable estimation of specific perturbations $p(A_1^{11}, E_A^X)$ in all predicted compounds comprising site A_1^{11} by the local average perturbation $\bar{p}_1(A_1^{11}) = 1/2\,(p(A_1^{11}, E_A^m) + p(A_1^{11}, E_A^M))$. The anterior environment of all the predicted compounds lies between those tested and the predictions are all based on a topological interpolation.

$\longrightarrow \quad p(A_1^{11}, E_A^m)$ or $p(A_1^{11}, E_A^M)$,

$\xrightarrow{\;***\;} \quad p(A_1^{11}, E_A^X) \simeq p(A_1^{11}, E_m^M)$

For each preference structure, the average perturbations $\bar{p}(s)$ associated with each site are valuated on the one hand in their minimal anterior environment and on the other hand in their maximal anterior environment. The minimal compounds enable us to

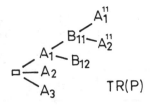

TR(P)

$\mathcal{P}_c$	STRUCTURAL VARIABLES	MINIMAL COMPOUNDS		MAXIMAL COMPOUNDS	
		THEORETICAL	AVAILABLE	THEORETICAL	AVAILABLE
	FO	▫			
	A_1	▪—●	1		
	A_2	▪⟨●	2		
	A_3	▪⟨●	3		
$\mathcal{P}_c^1$ $\mathcal{P}_c^2$	B_{11}	▪—●—●	4	▪⟨●	6
	B_{12}	▪—●⟨●	7	▫⟨○—○	
	A_1^{11}	▪—●—●—●	10	▫⟨○—○—○	12 13
	A_2^{11}	▪—●—●⟨●	14	▫⟨○—○—○	▪⟨●—●⟨● 15
	$A_2 * A_2^{11}$	▪⟨●—●⟨●	15		

Figure 5.21. Theoretical and available key populations

The level 1 theoretical key population, p_c^1, for a treatment which requires only primary sites, groups the twelve minimal and maximal compounds associated with the focus and with the primary sites of the trace. The level 1 available key population includes eleven compounds: eight whose structures are the same as those of the theoretical key compounds and three most closely resembling the theoretical ones. For example, the maximal compound associated with A_1^{11} is replaced by compounds **12** and **13**.

estimate the constant part of the perturbation, $p(s, E_A^m)$, regarding the introduction of site s in the presence of those which condition its existence. The maximal compounds enable us to estimate, in addition, the variable part $p(s * (E_A^M - E_A^m))$ reflecting the contributions of all the cosites between s and the variable part of its environment $(E_A^M - E_A^m)$ (cf. 2.2.1):

$$p(s, E_A^M) = p(s, E_A^m) + p(s * (E_A^M - E_A^m))$$

When the perturbations $p(s, E_A^M)$ and $p(s, E_A^m)$ associated with the minimal and maximal anterior environments are similar for each site, then the variable part of each perturbation is almost zero. There then exists a strong probability that no cosite influences the property, and the estimation of each specific perturbation $p(s, E_A^x)$ by the average perturbation $\bar{p}(s)_m^M$ is completely reliable for all the compounds of the proference. When the specific perturbations $p(s, E_A^M)$ and $p(s, E_A^m)$ are too far removed from each other, considering the experimental precision, then the minimal compounds associated with the cosites influencing property are introduced into the key population. They enable us to estimate the perturbations associated with these detected cosites, and the estimation is once again completely reliable for all the compounds of the proference.

Thus, the thirteen compounds of the theoretical key population provide a completely reliable estimate of the activity of the nine proference compounds whose structures are all anteriomorphs of the structures handled.

5.3.3.2 Modulated key population for extending the proference and optimizing its reliability

The design of the modulated key population or optimal experimental population to be treated comprises three main stages:

The theoretical level 2 key population, P_c^2, for a treatment which requires introducing level 2 cosites, includes thirteen compounds: the twelve of the level 1 theoretical key population plus the minimal compound **15** associated with the detected binary cosite $A_2 * A_2^{11}$. The level 2 available key population includes only the eleven compounds of the level 1 available key population, since compound **15** already belongs to this population.

Constructing the structural area to be explored – Determining the key population – Modulating the key population, given the constraints.

Constructing the structural area to be explored

The structural area is deduced by analyzing the results of the Topo-Information relationship. It includes the favorable activity environment E_F^* (10d) when the study is aimed at optimizing a series, the active environment E^* (1a) when the study is a mechanism search. It can be extended by introducing sites to determine or to specify the activity frontier F_A or the zero influence frontier F_0.

The goal of the glycuronic acid conjugation study is the search for a prediction law. The structural area to be explored includes all the sites of the studied environment, as they all influence the property. It is extended by introducing new sites in searching for the zero influence frontier F_0 (Figure 5.22).

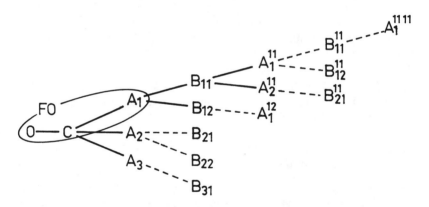

Figure 5.22. Structural area to be explored in searching for the zero influence frontier

The structural area to be explored groups those environment sites participating in activity, in this case all those of the area studied. When searching for a prediction law, it can be extended by introducing new sites whose aim is to determine the zero influence frontier F_0.

When initializing an experiment plan, this structural area is built by grouping the structures having an activity in the studied biological area (10d). It can eventually be extended by knowing the influence of some structural elements in similar series.

In the case of an a posteriori treatment, this area is defined by the trace of all the experimentally available structures.

Determining the key population

The key population includes the compounds relevant to the establishment of Topo-Information correlations, for obtaining a broader and more reliable preference. Those are the minimal and maximal compounds associated with the topochromatic sites of the trace of the structural field to be explored. However, in order to avoid too many syntheses, the compounds requested are limited to the minimal compounds during the elaboration phase. The maximal compounds are only requested for those sites belonging to the trace of the active field explored, – in this case, the whole studied field.

Thus, in this example, the key population includes the 13 minimal and maximal compounds associated with the sites of the active environment (Figure 5.21) for optimizing the reliability of the proference. It also includes the 8 minimal compounds associated with the new sites of the structural area to be explored in order to extend the proference (Figure 5.22).

In a posteriori treatment, this population includes all the minimal and maximal compounds associated with the trace sites. It corresponds to the theoretical key population defined in the previous paragraph.

Modulating the key population

This stage takes account of the constraints due to experimental difficulties (difficulty of synthesis – solubility – availability). These properties are experimentally known or deduced from Topo-Information relationships (10g, 10h, 29).

In a posteriori treatment, only the availability constraints are considered. An available key population is defined. It includes the experimental compounds most resembling the theoretical key ones. Thus, the available key population in this study includes 11 of the

13 experimental structures of table 5.1 (Figure 5.21). One minimal and three maximal key structures are missing. The absence of A_1^{11} maximal environment is partially compensated for by introducing structures 12 and 13. The absence of A_2^{11} maximal environment is partially compensated for by the presence of structure 15. Structures 5 and 11 are also available but do not intervene in the key population modulated by availability. They are used to check the preference retrospectively.

In a true experiment plan, the key population is screened by criteria of experimental possibilities. In our example, the trace to be explored includes 8 sites and cosites of the active environment and 8 new sites whose purpose is the search for the zero influence frontier. For the new sites, only the minimal environments are explored: the corresponding minimal compounds are probably apt to be synthesized and their activity measurable. The influence of the empty environment cannot be measured and compound 0 constituted by the focus with hydrogens is excluded from the modulated key population. Concerning the maximal environments of sites B_{12}, A_1^{11} and A_2^{11}, the difficulty of synthesis must be compared to the corresponding increase of preference reliability. One chooses among the experimentally accessible compounds those which most increase the completely reliable proference. Thus, knowledge of compounds marked $+0$ (Figure 5.23) is not advantageous, since it does not increase the completely reliable preference. However, knowledge of compound 31, marked $+3$, increases the completely reliable proference by three structures. Among these, two are experimentally known, 12 and 13, and thus constitute useless syntheses. Knowledge of the three theoretical key compounds 30, 31, 32 instead of the four non-key experimental structures 5, 11, 12 and 13 would, in this case, have extended the completely reliable proference from 2 to 9 (Figure 5.23). This shows the importance of carrying out an interactive strategy from the beginning, in order to obtain the best compromise between the predictive power of the correlation and experimental efficiency.

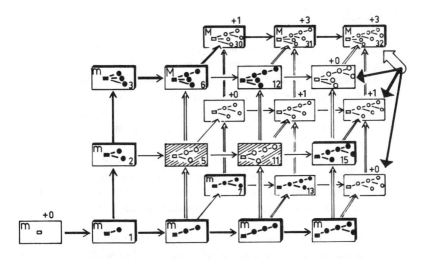

Figure 5.23. Modulated key population for optimizing proference reliability

If, instead of the two non-key experimental structures **12** and **13**, we know the three theoretical key compounds **30**, **31**, **32**, the completely reliable proference will increase from 2 to 9. The index on the upper right of each structure indicates the increase of completely reliable proference when this compound is available. Knowing compound **32** (marked +3 and with a white arrow) makes possible a completely reliable estimation of the behaviour of the three structures marked with a black arrow. In experiment planning, the compounds requested, chosen among those experimentally available, are those which most raise prediction reliability.

	theoretical	available
key structures		
completely reliable predicted structures		

5.4 Conclusion

The DARC/PELCO method proves itself to be a very flexible and precise tool for estimating biological activity of compounds from the structural elements which characterize them. In order to find the best model, one detects intrinsic regularities of phenomena evolving along the ordered pathways of structural filiations. Inference tools proceed by heuristic modulation of the initial representation space of experimental data. The Structure-Activity relationships obtained constitute a simple law whose precision is comparable to that of measurements. The regularities detected suggest a structural interpretation. The predictive capacity is both broad and reliable and its validity beyond the field explored can be inferred.

By inserting a structure into a series, called a hyperstructure, we define the concept of structural interpolation which engenders fundamental tools. Selecting the optimal key population ensures maximal structural interpolation, given the experimental difficulties. It guarantees optimization of the activity perturbations and of the prediction area. The prediction reliability of each new structure is determined according to its location in the hyperstructure with respect to the experimental structures. Experiment planning helps to extend the prediction area and optimize its reliability.

The experimental population which illustrates this presentation was chosen purposely, restricted but representative, comprising both gaps and redundancies. However, the missing key structures are less numerous than the useless structures. Since the optimal population is not available, interpolation is difficult, whereas experimental knowledge of the missing key structures would have engendered a completely reliable proference.

The DARC/PELCO method is now largely automated. A computerized version with a graphic interactive interface will be presented in a forthcoming paper.

5.5 References

1. a) J.E. Dubois, D. Laurent and H. Viellard C.R. Acad. Sci. Paris, 1967, 264C, 1019;

b) J.E. Dubois, D. Laurent and A. Aranda, J. Chim. Phys., 1973, 11–12, 1608 and 1616.

2. a) J.E. Dubois, D. Laurent and H. Viellard, C.R. Acad. Sci. Paris, 1966, 263C, 764.

   b) J.E. Dubois, DARC System in Chemistry, in "Computer Representation and Manipulation of Chemical Information", W.T. Wipke, S. Heller, R. Fellmann and E. Hyde, Eds., Wiley, New York, 1974, 239; J.E. Dubois, Isr. J. Chem., 1975, 14, 17.

   c) J.E. Dubois and Y. Sobel, J. Chem. Inf. Comput. Sci., 1985, 25, 326.

3. a) C. Mercier, V. Fabart, Y. Sobel and J.E. Dubois, J. Med. Chem., 1991, 34, 934.

   b) C. Mercier, Y. Sobel and J.E. Dubois, in "QSAR: Rational approaches to the design of bioactive compounds", C. Silipo and A. Vittoria, Eds, Elsevier, Oxford, 1991, 533.

   c) C. Mercier, V. Fabart, Y. Sobel and J.E. Dubois, "Les systèmes d'information en chimie", S.F.D.I.C., 1990, 28.

4. a) J.E. Dubois and A. Panaye, Tetrahedron Letters, 1969, 1501 and 3275; J.E. Dubois, A. Panaye and J. MacPhee, C.R. Acad. Sci. Paris, 1975, 280C, 411;

   b) A. Panaye, J.A. MacPhee and J.E. Dubois, Tetrahedron, 1980, 36, 759;

   c) J.E. Dubois, Pure Appl. Chem., 1981, 53, 1313;

   d) J.E. Dubois and J. Chrétien, J. Chromatograph. Sci., 1974, 12, 811.

5. S.M. Free and J.W. Wilson, J. Med. Chem., 1964, 7, 195.

6. L.B. Kier and L.H. Hall, Molecular Connectivity in Chemistry on Drug Research, in "Medicinal Chemistry", Vol. 14, G. de Stevens, Ed., Academic Press, London, 1976.

7. C. Hansch and T. Fujita, J. Am. Chem. Soc., 1964, 86, 1616; C. Hansch, Quantitative Structure-Activity Relationships in Drug Design, in "Drug Design", Vol. 1, E.J. Ariens, Ed., Academic Press, London, 1971, 271–342.

8. J.E. Dubois and H. Herzog, J. Chem. Soc., Chem. Commun., 1972, 932.

9. a) J.E. Dubois and A. Massat, C.R. Acad. Sci. Paris, 1967, 265C, 757;

   b) J.E. Dubois, Pure Appl. Chem., 1977, 49, 1029;

c) J.E. Dubois and M. Carabedian, Org. Magn. Reson., 1980, 14, 264.

10. a) J.E. Dubois, Bull. Chim. Ther., 1972, 1, 65;

b) A. Aranda, C.R. Acad. Sci. Paris, 1973, 276C, 1301;

c) J.E. Dubois, in "Man and Computer", North Holland Publ. Co. Inc., New York, N.Y., 1974, 309;

d) J.E. Dubois, D. Laurent, P. Bost, S. Chambaud and C. Mercier, Eur. J. Med. Chem., 1976, 11, 225;

e) B. Duperray, M. Chastrette, M.C. Makabeh and H. Pacheco, Eur. J. Med. Chem., 1976, 11, 323 and 433;

f) C. Mercier, Y. Sobel and J.E. Dubois, Eur. M. Med. Chem., 1981, 16, 473;

g) Y. Sobel, C. Mercier and J.E. Dubois, Eur. J. Med. Chem., 1981, 16, 477.

h) J.E. Dubois and C. Mercier, Livre jubilaire en l'honneur du Professeur R. Truhaut, 1984, pp. 332.

i) J.E. Dubois, C. Mercier and A. Panaye, Act. Pharm. Yougosl., 1986, 36, 135.

11. a) C. Mercier and G. Trouiller in "Quantitative Structure-Activity Relationships in Drug Design", J.L. Fauchère, Ed., Alan R. Liss Inc., 1989, pp. 203;

b) C. Mercier, G. Trouiller and J.E. Dubois, Quant. Struct. Act. Relat., 1990, 9, 88.

12. I.A. Kamil, J.N. Smith and R.T. Williams, Biochem. J., 1953, 53, 129.

13. C. Hansch, E.J. Lien and F. Helmer, Arch. Biochem, Biophys., 1968, 128, 319.

14. C. Mercier and J.E. Dubois, Eur. J. Med. Chem., 1979, 14, 415.

15. G. Sicouri, Y. Sobel, R. Picchiottino and J.E. Dubois, "The role of Data in Scientific Progress", P.S. Glaeser, Ed., Elsevier Science Publishers B.V., North Holland, 1985.

16. J.E. Dubois, F. Hennequin and M. Chastrette, Bull. Soc. Chim. France, 1966, 11, 3568; J.E. Dubois and H. Viellard, Bull. Soc. Chim. France, 1968, 900, 905 and 913; J.E. Dubois, Ordered Chromatic and Limited Environment Concept, in "Chemical Applications of Graph Theory", F. Harary and A.T. Balaban, Eds., Academic Press, London, 1976.

17. J.E. Dubois, D. Laurent, A. Panaye and Y. Sobel, C.R. Acad. Sci. Paris, 1975, 280C, 851 and 281C, 687.

18. J.E. Dubois and A. Panaye, Bull. Soc. Chim. France, 1975, 2100.

19. J.E. Dubois, M.J. Alliot and H. Viellard, C.R. Acad. Sci., Paris, 1970, 271C, 1412; J.E. Dubois, M.J. Alliot and A. Panaye, ibid, 1971, 273C, 224; J.E. Dubois, A. Panaye and P. Cayzergues, ibid, 1980, 290C, 429.

20. J.E. Dubois, C. Mercier and Y. Sobel, C.R. Acad. Sci., Paris, 1979, 289C, 89.

21. J.E. Dubois, Y. Sobel and C. Mercier, C.R. Acad. Sci., Paris, 1981, serie II, 292, 783.

22. A. Cammarata and S.Y. Yau, J. Med. Chem., 1970, 13, 93.

23. L.H. Hall and L.B. Kier, Eur. J. Med. Chem., 1978, 31, 89.

24. W.R. Smithfield and W.P. Purcell, J. Pharm. Sci., 1967, 56, 577.

25. C. Hansch and S.H. Unger, J. Med. Chem., 1973, 16, 217.

26. J.G. Topliss, J. Med. Chem., 1972, 15, 1006.

27. J.E. Dubois and D. Laurent, C.R. Acad. Sci. Paris, 1969, 268A, 405.

28. J. Johnston, Econometric Methods, McGraw-Hill Book Co. Inc., London, 1963; N.R. Draper and H. Smith, Applied Regression Analysis, John Wiley and Sons Inc., London, 1966; J. Lellouch and P. Lazar, Methodes statistiques en experimentation biologique, Flammarion Medecine Sciences, Paris, 1974.

29. J.E. Dubois, A. Panaye and C. Lion, Nouv, J. Chim., 1981, 5, 371.

INDEX

INDEX